JACKSON
SYSTEM
DEVELOPMENT

JACKSON SYSTEM DEVELOPMENT

Alistair Sutcliffe
UMIST

PRENTICE HALL

NEW YORK   LONDON   TORONTO   SYDNEY   TOKYO

First published 1988 by
Prentice Hall International (UK) Ltd,
66 Wood Lane End, Hemel Hempstead,
Hertfordshire, HP2 4RG
A division of
Simon & Schuster International Group

Printed and bound in Great Britain by
A. Wheaton & Co. Ltd, Exeter.

LIBRARY OF CONGRESS CATALOGING–IN–PUBLICATION DATA

Sutcliffe, Alistair, 1951–
   Jackson system development / Alistair Sutcliffe.
       p.   cm.
   Bibliography: p.
   Includes index.
   ISBN 0–13–508136–X
   1. Jackson system development method.
   2. Electronic data processing – Structured techniques.
   I. Title.
   QA76.9.S88S88 1988
   004.2'1 — dc19   87–22323

BRITISH LIBRARY CATALOGUING IN PUBLICATION DATA

Sutcliffe, Alistair
   Jackson system development.
   1. Electronic digital computers –
   Programming   2. System analysis
   I. Title
   004.2'1   QA76.6

   ISBN 0–13–508136–X
   ISBN 0–13–508128–9 Pbk

---

1 2 3 4 5   92 91 90 89 88

ISBN 0-13-508136-X
ISBN 0-13-508128-9 PBK

To my father and mother

CONTENTS

PREFACE

This book aims to explain the Jackson System Development (JSD) method to those in education and industry who are interested specifically in JSD and more generally in systems analysis and design. JSD has come a long way since its birth in 1983 when Michael Jackson's definitive book *System development* was published. Since then further aspects of JSD have been added in John Cameron's *JSD and JSP: the Jackson approach to software development* and the method has evolved as experience in its use has increased. JSD has grown on the firm foundations of Jackson Structured Programming (JSP), which since its definition in 1975 has become the *de facto* standard of structured program design in many commercial organizations. Based on the foundation of JSP, JSD is becoming one of the leading contenders in the current generation of structured systems analysis and design methods.

While JSD is not alone in the school of structured analysis and design methods, it does offer a clear approach to the problems of systems development based on sound principles. When methods are classified in categories representing their emphasis and founding principles many show close relationships and eclectic development, i.e. they have borrowed ideas from elsewhere. JSD stands out in having a clear approach of its own, however, and merits study for its differences from other methods as well as for its inherent qualities as a practical development method.

The information in this book has been derived from the two previous publications on JSD, from notes and other course material kindly provided by Michael Jackson Systems Limited and from a variety of seminars and presentations on the method. It is also based on my experiences of teaching the method to undergraduates at UMIST and research into the problems inherent in systems development. While the explanation of the method is, I trust, faithful to the method, much of the material, in particular the case studies, is of my own making. It is inevitable that some individual interpretation creeps into a text of this nature but I do not regard this as detrimental. Accordingly I have not tried to eliminate such influences; my aim is not to reiterate JSD but to interpret it as well.

It is a pleasure to acknowledge the generous help of Michael Jackson Systems Limited in providing material for preparation of this book and the efforts of John Cameron and Ashley McNeile whose patient and critical evaluation of the manuscript improved the description of JSD and kept its interpretation on the correct path. I take the responsibility for any errors which escaped their attention. Finally I am grateful for Michael Jackson's permission to use material from his original book on the method and for his support in writing this book.

A. SUTCLIFFE

1
INTRODUCTION TO JSD

1.1 THE BACKGROUND OF JSD

Systems analysis and design have been progressing through the 'structured revolution' which aims to replace traditional system development techniques with more formal methods for building reliable systems. The need for this change has been prompted by the poor reliability of system designs and the growing nightmare of modifying systems – a situation which has led many software developers to question *ad hoc* approaches to building systems. The combination of spending more and more time trying to maintain unmaintainable systems and trying to debug unreliable behaviour in new systems is imposing a near-intolerable burden on many data processing departments.

This book aims to describe one method, Jackson System Development (JSD), which has been proposed as a potential solution to these problems. Since its launch in 1983 in Michael Jackson's book *System development*, JSD has gained acceptance in commercial data processing departments, often building on its predecessor Jackson Structured Programming (JSP). JSD is not without competition, however, so this book has been written not only to explain JSD to analysts and students who wish to use it, but also to set JSD in the wider perspective of structured system development methods.

Structured methods

Poor system design has resulted in the problems of error-prone and unmaintainable systems. Attempts to alleviate these problems have given rise to several different methods, all of which claim to be structured; this can leave the layman in a state of confusion when bombarded with claims and counter-claims for the various methods.

Firstly we should examine what 'structured' means, or rather how the term has been interpreted by its proponents. Most structured methods break the process of systems analysis and design down into a series of steps. These steps consist of a set of heuristics, that is, logical advice about how to proceed with design – rather as a recipe guides a chef through the steps of creating a meal. Authors of systems design methods may be offended to see their craft compared with cooking; however, both have a heuristic basis, and offer sound advice which guides the practitioner towards a successful solution while not guaranteeing that the results will be correct or perfect.

In addition to the procedures, most structured techniques make extensive use of diagrams to represent facts about systems. These diagrams are used to create a model which shows an abstract view of a system composed of processes, data and control of activity. The model may be supplemented with some type of specification language to describe activities in a concise form. Diagrams add clarity to specifications and can overcome the ambiguities inherent within English language descriptions. Specifications tend to be shorter, neater and less verbose.

The process of structured design has underlying principles upon which the method is based. These principles will be derived from other areas of computer science such as program design, data modelling and computational theory. But there the similarities of structured methods end.

Differences appear in the emphasis methods place on different components in systems; for instance, some pay more attention to data analysis while others concentrate on functional aspects. Further divergence appears in the phases of the system's life cycle covered by each method; some give comprehensive cover to the analysis phase with little advice on design and the reverse case can also be found (Maddison *et al.* 1983). Finally methods vary in the formality and rigour of their techniques, that is, how far their approaches to system specification and design are founded on mathematical principles which can be verified.

None of the present 'first generation' of structured methods is rigorous in the true sense that their specifications are formulated mathematically. Second-generation methods are still at the research stage and even if the theory of mathematically correct specification is created, the lack of an acceptable user interface to hide complicated mathematics from users will hinder their acceptance. This leaves the software developer with the aim of finding a method which covers most of the system's life cycle, pays attention to data and the processing parts of systems and has some formality so that specifications can at least be verified. This book aims to put Jackson System Development into that perspective.

Jackson System Development is a system specification and design method which covers the system's life cycle from analysis through to implementation. It has undergone considerable development since it was first described; I shall describe the method as it is now, although with some notes to previous versions so readers can refer to previous texts on the subject (Jackson 1983, Cameron 1983). It differs in many respects from other methods in the 'structured school' and has to be placed on its own in comparative schemas of methods. On the other hand JSD is not too adventurous and has its foundations firmly based in the realities of practical systems analysis and design.

JSD has a common aim with all structured methods to create more reliable and maintainable systems, partly by better design and partly by creating more concise, unambiguous and readable specifications. Jackson's emphasis on modelling the problem environment is different from other methods, however, and it is this clear view of the problem which in Jackson's view enables a stable and reliable system to be built to solve the problem. The unique approach of JSD, which places it apart from the mainstream of functional top-down approaches to systems analysis, lies in the concepts of modelling and time ordering (Maddison *et al.* 1983, Olle *et al.* 1986). Although hints of such concepts may be found elsewhere, no other method makes them as explicit as JSD does.

1.2 OVERVIEW OF JSD

JSD is organized in three separate stages which guide the analyst through the systems development process. Each stage has a set of activities with clear start and end points (this helps the analyst using the method) and facilitates project control as deliverables can be defined for each stage. The three stages can be outlined briefly as follows.

(a) *Modelling stage*. A description is made of the real world problem and the important actions within the system are identified. This is followed by analysis of the major structures within the system, called *entities* in JSD. This stage is described further in Chapters 2 and 3.

(b) *Network stage.* The system is developed as a series of subsystems. First the major structures are taken from the modelling stage and input and outputs are added; this is followed by the analysis of the output subsystem which provides information, and then of the input subsystem which handles the user interface and validation. Chapters 4 and 5 deal with this stage.

(c) *Implementation stage.* In this stage the logical system specification, which is viewed as a network of concurrently communicating processes, is transformed into a sequential design by the technique of scheduling. This is followed by further detailed design and coding. More particulars are given in Chapter 6.

JSD begins by analysing the major system structures which are important to create a model of the system problem, the entities. Then these structures are connected together to create a network model of the system, while at the same time the design is elaborated by addition of other processes to create output, and to handle input messages and user interaction. The essence, as is explained in more depth later, is to create a system model of reality first and then to add the functionality.

Design considerations are delayed to the last stage in JSD. Until then no assumptions are made about the machine the system is to run on and the logical system view is one of many concurrent communicating processes. In the last stage this view is transformed into a physical design, a step referred to as 'implementation' in JSD. Design detail may be added in earlier stages; however, the remaining specification details are added during implementation before coding can begin.

The stages of JSD underlie its approach which is explained in more depth in the philosophy of the method.*

1.3 PHILOSOPHY OF JSD

The underlying philosophy of JSD is deceptively simple and may be summarized in the following set of principles:

---

* Readers of previous versions of JSD will remember that the original version had six steps. The mapping is approximately as follows: Entity action and Entity structure to the Modelling stage; Initial model, Function and System timing steps to the Network stage; and Implementation stays as it was.

- Before building systems, understand them.
- To understand systems, make a model.
- Systems are all about changing things, therefore model the way things change.

Such a reductionist view of JSD may sidestep the complexities of the method but it does provide an initial reference point for a vital part of JSD. The idea of modelling the system is very important in JSD in two senses. First, a model is created, in order to understand the system. Although this may be done in other methods (for instance, the data flow diagrams of structured analysis are a kind of model – see De Marco 1978), in JSD more emphasis is placed on this activity. Secondly, there is the subject matter to be modelled.

JSD guides the analyst to model the system's basic structure rather than some of its more obvious manifestations as functions, which Jackson separates from the basic system model. Other methods, in particular the structured analysis and design school, concentrate on functions. Jackson regards this approach as flawed because it leads to inaccurate modelling of the system structure and the problem it aims to solve.

In contrast JSD focuses on the system problem. The problem can be considered as a situation which has to be changed by the system in order to meet one or more goals of the organization. This view is inherent in JSD's approach to functionality, which is not modelled until the later stages of JSD. While it is difficult to be exact about 'basic structures' of systems the effect is to consider the system as a problem which has to be solved. By considering the problem itself rather than the current solutions (i.e. functions of the system) JSD aims to get at deeper structures than functional methods do.

Jackson places emphasis on correct modelling because he claims that only the use of this approach can build accurate and reliable systems which reflect the realities of the real world. Problems can be viewed as an undesirable state of affairs which have to be changed by the system's activity. Modelling that activity focuses on how things change over time. Since systems are all about change it is important to model how and when that change takes place. JSD maintains that if our view of how things change is correct and the events or triggers that cause change to happen have been accurately recorded, then the computer system should behave correctly. This view in inherent in JSD time ordering.

1.4 STRUCTURED DESIGN AND JSD

A brief diversion at this point is taken to place JSD in the perspective of other contending structured design methods. A complete review is beyond the scope of this book; more detail can be found in Maddison *et al.* (1983) and Olle *et al.* (1986).

All structured systems development methods create models of the world, but the models they create differ. Broadly they differ in the type of information they capture and the level of abstraction at which they specify. The level of abstraction is a gradient from physical machine detail to logical designs to conceptual models of the problem and is usually divided into three levels: conceptual, logical and physical. The type of information falls into three submodels which combined describe most qualities of a system. These are:

- The process communication model, which specifies the system's activity and data communications between processes.
- The data relationship model, specifying the data structures and their relationships.
- The state–event model, which views activity in terms of a series of transformational steps (events) between periods of inactivity (states).

All three models are related and to an extent complementary. A method should contain all three and provide for different levels of abstraction for each model as analysis proceeds from the conceptual through the logical level to design at the physical level.

Most of JSD's rivals fall into the functional school of methods which concentrate on the process communication model. Most influential of these is Structured Analysis/Structured Design (De Marco 1978, Yourdon and Constantine 1977) which views systems as a network of functional processes each of which achieves a goal. No attempt is made in the original method definition to model the underlying system structure.

More recent methods have emerged which are more eclectic in their background. These methods of which SSADM (or LBMS–SDM in its commercial version, Learmonth and Burchett Management Systems 1986) and Information Engineering (Macdonald 1986) are important examples, have acquired process communication models in the form of data flow diagrams; from research on data analysis they have acquired entity relationship models, and state transition models have been added to complete the picture. While these methods are undoubtedly comprehensive they may be criticized as being cumbersome. The multiplicity of models leads to duplication of effort in analysis, consequently creating potential problems of inconsistency and redundancy in the specification.

JSD, in contrast, places more importance on state–event modelling, although it does incorporate process communication modelling in the network stage. Unlike other methods JSD does not analyse all the models separately, but integrates this activity and uses the implicit relationships between models to derive further information. Thus, as we shall discover, JSD does not model data relationships directly but obtains this information by inference from the process communication model. It is the clear emphasis on modelling and the integration of the JSD specification which marks it out from other methods. Another virtue of JSD is its support for different levels of abstraction. The different stages of JSD concord with the idea of levels of abstraction, and the emphasis JSD places on delaying consideration of physical design details until the implementation stage gives a clean division between the logical and physical levels in specification.

1.5 EVOLUTION OF JSD

Readers may be interested to gather some idea of where JSD fits into the broader perspective of computer science research on systems and program design. Michael Jackson's ideas come from mathematical and theoretical concepts about program design, which he translated into a practical method which started as Jackson Structured Programming (JSP – Jackson 1975) and later grew into JSD (Jackson 1983).

The first point to make clear is that these form a quite radical bridge between the formality of mathematical aspects of computer science and the practical aspects of commercial systems development. In common with most innovatory ideas JSD draws on previous work, but its synthesis is something new.

Academic computer scientists have been concerned for a long time with questions about how to specify a 'good' program structure and what should be the contents of that structure and its subcomponents. For academics the idea of 'goodness' has been synonymous with mathematical rigour, that is, the ability to be able to specify program instructions and their sequence of operation in an algebraic or grammar-like formalism so that the program's behaviour can be proved to be correct or at least verified as being well designed.

This quest has given rise to two approaches. Program provers have attempted to construct programs so that all possible processing pathways (different sequences of execution of instructions) can be checked out, thus ensuring the program will function correctly in all circumstances. On the other hand, program structuralists have attempted to define better subcomponents of programs and by doing so to limit the activity of a program to a more exact definition of what it can and cannot do. JSD owes its intellectual roots to both schools, perhaps leaning a little more closely to the latter.

The concept of object-oriented programming has grown out of the structural school, a prime example being the Smalltalk* language from Xerox (Goldberg and Robson 1983). This concept is based on the idea that programs should be built around models of things in the real world. The aim is that programs should reflect the behaviour of real world objects rather than their structures being designed for computer processing. JSD fits the object-oriented approach with its emphasis on modelling the real world and then building programs based upon that model.

Prominent in JSD is the view of a system composed of communicating yet separate processes. This is derived from schools of thought which have proposed that programming structures should be more clearly based on input and output rather than on procedural code, as exemplified by Hoare's (1978) communicating sequential processes and Kahn and MacQueen's (1980) networks of parallel processes. Both of these approaches consider that systems are composed of smaller units which pass data between each other and that the overall behaviour of the system is governed by the mechanism of communication. This view, incidentally, anticipates future architectures in which computers may be constructed of many interconnected processors, in which case the mapping between process design and executing hardware may be very close indeed. The notion of transformation of concurrent systems into sequential architectures is based on the work of Burstall and Darlington (1975).

The other core concept in JSD and JSP is that of time ordering. This can be traced back from actions to events, which form the boundaries of actions: one event starts an action and another stops it. From events

---

* Smalltalk™ is a trademark of the Xerox Corporation.

it is a short leap to state–event transition diagrams and establishing programs as a series of steps which require specific pre- and post-conditions for correct execution. Academics of the program-proving school have been concerned with pre- and post-condition specification of program execution, the idea being that if all the triggering conditions can be accurately specified then it should be possible to predict when and why each bit of code will execute. Proponents of this school have led to the notions of structured programming, 'stepwise refinement' and more recently work on program proving. The reader may consult Gries (1981) for a summary; if the detail fascinates, Dijkstra (1976) and Hoare (1978) are definitive references.

JSD is not directly based on this school, but its emphasis on time in synchronizing processes and some of the process structure diagram notation (such as exclusive multipart conditions and iterations with explicit termination conditions) do owe their heritage to work in this area. The above exploration of some of the research issues in computer science should, at least, establish that JSD has drawn upon academic concepts in computer science and it may therefore justly claim to have an intellectually sound base. It is not the intention of this book to explain that base in detail; the references cited above should give the reader plenty of material with which to explore the literature.

1.6 JSD AND JSP

As we have seen, JSD is not the first method proposed by Michael Jackson. Since its appearance in 1975 Jackson Structured Programming has gained widespread acceptance and is practised in many organizations. So what is the relationship between JSP and JSD? The answer is that although JSD supersedes JSP, parts of JSP are preserved within JSD. JSD is a new method which covers the development life cycle from analysis to programming, and sidesteps much of the complexity of JSP with a simpler view of program specification and construction.

Having said that, it should come as some comfort to users of JSP that many of the concepts live on in JSD. Backtracking, program inversion and structure diagrams are all alive and well; what has vanished is the emphasis on data modelling, which is replaced by process modelling. As a result the emphasis on correspondence has gone and the program structure is modelled directly from analysis of the problem specification.

However, classic JSP data modelling and correspondence can be used to develop the information subsystem within JSD when process structures are highly dependent on data structures.

The later steps of JSP (elementary operations, assigning actions and conditions) are still present; so is schematic logic, except that JSD renames it *structure text*. Many of the ideas inherent in JSP are carried forward into JSD. The approach has changed but the underlying concepts are the same. With that in mind we should move on to examine what those concepts are.

1.7 SUMMARY

JSD is a structured method of software development which aims to produce more reliable and maintainable systems. It is divided into three stages to guide the analyst through the phases of development. The basic principle of JSD is modelling the real world system and thereby creating a computer system which can respond correctly to that reality. Modelling is done by finding the events of importance, called actions in JSD. Actions are modelled in time-dependent sequences.

This approach sets JSD apart from other structured system development methods, which emphasize different models and approaches to system specification. Although JSD and other methods may collect the same information about a system, JSD does so in a different manner which, it may be argued, avoids inconsistencies and ambiguities. The principles of JSD are based on research in program design methods which has created concepts of object-oriented programming, concurrent systems and methods of proving program behaviour. JSD does not supersede its predecessor, JSP, but complements it by extending the method to encompass problems of systems development rather than just program design.

REFERENCES

Burstall, R.M. and Darlington, J. 1975. 'Some transformations for developing recursive programs'. *Sigplan Notices* **10** (6): 465–72.

Cameron, J.R. 1983. *JSP and JSD: the Jackson approach to software development*. Los Angeles: IEEE Computer Society Press.

De Marco, T. 1978. *Structured Analysis and System Specification*. New York: Yourdon Press.

Dijkstra, E.W. 1976. *A Discipline of Programming*. Englewood Cliffs, NJ: Prentice Hall.

Goldberg, A. and Robson, D. 1983. *Smalltalk–80: the language and its implementation*. Reading, MA: Addison Wesley.

Gries, D. 1981. *The Science of Programming*. New York: Springer.

Hoare, C.A.R. 1978. 'Communicating sequential processes'. *Communications of the ACM* **21** (8): 666–67.

Jackson, M.A. 1975. *Principles of Program Design*. London: Academic Press.

Jackson, M.A. 1983. *System Development*. Hemel Hempstead: Prentice Hall.

Kahn, G. and MacQueen, D.B. 1981. 'Coroutines and networks of parallel processes'. In Bergland, G.D. and Gordon, R.D. (eds), *Software Design Strategies*, Los Angeles: IEEE Computer Society Press.

Learmonth and Burchett Management Systems 1986. *Structured System Analysis and Design, version 3*. London: LBMS.

Macdonald, I.G. 1986. 'Information engineering: an improved, automatable methodology for the design of data sharing systems.' In Olle *et al.* (eds) *Information Systems Design Methodologies*, 173–224. Amsterdam: North Holland.

Maddison, R.N., Baker, G.J., Bhabuta, L., Fitzgerald, G., Hindle, K., Song, J.H.T., Stokes N. and Wood J.R.G. 1983. *Information Systems Methodologies.*, BCS/Wiley Heyden, Chichester.

Olle, T.W., Sol, H.G. and Verrijn-Stuart, A.A. (eds) 1986. *Information Systems Design Methodologies: Improving the practice*. Amsterdam: North Holland.

Yourdon, E.E. and Constantine, L.L. 1977. *Structured Design: fundamentals of a discipline of computer program and system design*. New York: Yourdon Press.

2

BACKGROUND AND CONCEPTS

In this chapter some of the underlying concepts of JSD are explored. First the JSD idea of system models is introduced and compared with the 'top-down' functional decomposition approach. Then time ordering is examined to discover why it is important in JSD and how timing is specified and recorded. This leads into discussion of long-running processes, an important JSD notion which forms the basis for the analysis of entities. Entities are the components of the system which are essential to the model; the JSD approaches to system structures and discovering entities are described. Entities and their subcomponents called actions are defined and illustrated, with their recording in a diagrammatic representation called process structure diagrams (PSDs) and the equivalent text representation, structure text.

2.1 THE JSD MODEL

A model is a representation in an abstract form of something in the real world. A model railway is a realistic model of the trains with which children can play. Inside signal boxes another more abstract model is used by the signalman to represent railway lines and trains on them. This model has a more serious purpose: it allows the signalman to visualize and control all the train movements in a station. We make and use models in many disciplines: generals use models for war games, chess is an abstract model of strategic warfare, and architects build models of new buildings. In all cases the model is built to represent something more manageably, so that it can be understood and manipulated. Computer systems are also amenable to this treatment (see Figure 2.1).

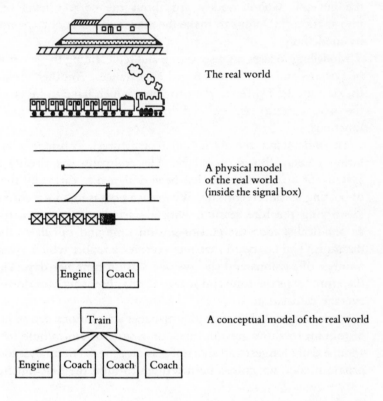

Figure 2.1 The JSD view of modelling: JSD creates conceptual models of real worlds.

JSD models systems to get a better view of what is happening in terms of activities and structures; the point of getting a more detailed and clear view of systems is to help the analyst and user understand systems behaviour and improve it. A good analogy for the analyst's problem is a general with a battlefield model: only by visualizing a battle with a model can the general see how the battle is going and direct it. Without the model both general and analyst remain in the dark and can only guess at what may be happening.

To be worthwhile, modelling must be accurate and exhaustive enough to describe the important things that are happening. An inaccurate model which does not show the enemy under camouflage netting is no good to the general; neither is a system model with hidden details. A battlefield model does not need to show details of where the soldiers' uniforms were made however – this detail is irrelevant to the general. Good models are about the correct level of relevant information. JSD aims to make the relevant facts about systems explicit by modelling.

Modelling focuses on discovering and representing the 'inner structures' of systems and not their external appearances. Another view of this is that the model expresses the problems which have to be tackled (with the new system) as opposed to its current solution, i.e. a set of functions.

To understand the term 'inner structures', consider a system for hiring pleasure boats on a lake. The company has already bought a system for boat hire which has been designed to carry out the function of boating session recording. When each boat is hired a record is created identifying the hire session, with a start time; on return the system automatically adds the end-of-session time and calculates the session duration. The boatyard manager receives a report which gives him the number of sessions and the average session time per day. The system therefore totals the time and sessions and subsequently performs the day average calculation.

So far so good, but soon the manager's ambitions grow; he wants a histogram to show session times on a scale of ten-minute increments, and the daily longest and shortest session times. The analyst looks at the program code which has been written to fulfil the first function:*

```
Session prog
    Init counters
    Get first session record
    Do while session records to be processed
        If start record
            Create new session
            Stamp with start time
        Else
            Match end record to start record
            If no match
                Signal error print bad session
```

* I am indebted to Michael Jackson for the original example.

```
            Else
                Calculate session time
                Add session time to total
                Increment session total
            End-if
        End-if
        Read next record
    End-while
    Print numbers of sessions
    Calculate average session time
    Print average session time
End session prog
```

The analyst concludes that the only way to implement the modification is to write each session duration to a separate file and then write another program to analyse session duration. The analyst is forced into this course of action because the first program does not preserve individual boating session times.

So far the analyst has survived. But then boating sessions become more frequent and start to overlap, which can create a large number of start records with no end records. This leads to a mismatch, so he has to redesign the system with a larger buffer and run in semibatch mode when he can reasonably expect most starts and stops to be present in the buffer (i.e. at lunchtime). Unfortunately this modification cannot cope with busy lunchtimes. This upsets the session time program, which expects only completed sessions to be transmitted to the session time analysis file. He fudges the issue by discarding unmatched sessions.

No sooner has the harassed analyst recovered from this problem than the manager becomes interested in the idle time of his boats. He suspects that canoes are more popular than rowing boats and wants a report of idle time by boat type. The analyst looks at his program (which doesn't even record the boat type) and applies for another job.

In this example the problems and modifications have accumulated until the system bears little resemblance to what was originally implemented and our beleaguered analyst decides that it has to be completely rewritten. His task has been made difficult because the original specification was modelled on functions, i.e. recording total time and sessions. The system had to be extended by addition of totally separate programs for new functions and this strategy soon became unworkable. If the analyst had investigated the underlying structure of the system (the nature of boating sessions) in the first place these problems would not have arisen.

JSD approaches this problem by modelling what happens in the real world, in this case what happens when people go boating. If the behaviour of the system is observed, recorded and modelled accurately then we have a correct picture of it. Information is going to be required about the system and its activity, so if our model is correct then we should have the information to hand rather than having to 'bolt on' bits to the system until it becomes unmanageable.

In the boating example the analyst should have modelled how boating sessions happen in reality. For instance, sessions may be started and finished, but may also be started but not finished (bad debts when dishonest customers leave the boat and do not pay) and may be finished but apparently not started (when the boatman forgets to key in a start record). Calculation of the day averages would be delayed until the end of the day's processing. Because the program would be explicitly based on boating sessions, modifications such as the requirement for a histogram of session time would present no problem because information about each session in the model is available.

The assertion is that system models are more stable than functions are. Many systems are built on functional concepts whereby the system is viewed as a set of procedures and tasks which are implemented to meet specific short-term goals, e.g. calculating an insurance premium, validating a customer's order or producing a sales report. Functions report some aspects of a system's behaviour but tell us little about the underlying causes of that behaviour; this type of analysis, therefore, resembles treating the symptoms of a disease rather than the cause. It takes time to 'think Jackson' and to ask questions about underlying processes rather than functions, but this does produce rewards in two ways:

■ The system's behaviour is modelled and made explicit, so helping understanding of what the system's purpose is.
■ Modelling systems helps to expose their deficiencies so that the analyst can make constructive suggestions about how they may be improved.

Models are not of course immune to change; the boatyard owner might decide to go in for selling ice cream, but that is an extension of his business which is a major step and a new subsystem in its own right. Functional changes are likely to be more frequent, and JSD anticipates this by separating functional processing from 'main' processing in the system model. The changeable bits of the system are separated from the basic processing; this should minimize the chance of system maintenance upsetting the original design and creating unforeseen errors.

2.2 ENTITIES AND ACTIONS

To model systems JSD employs entities and actions. These terms are more formally defined in the Glossary; however, useful working definitions for the concepts can be given as follows.

- *Entity*. An object of interest in the system which will undergo or cause change during the system's activity.
- *Action*. Event which happens to an entity.

It is important to distinguish between the JSD meaning of these terms and that of others. JSD entities are dynamic and active, which means that they are composed of actions and either effect or respond to changes in the real world. Database entities, in contrast, are static; they are changed by things in the system but do not themselves cause any change. A database entity merely records what has happened, whereas JSD entities cause and respond to events that happen. The two concepts are therefore complementary; the results of a JSD entity's actions will be recorded in a database entity structure. For example, a JSD Customer entity places, amends and cancels an order and the status of the Customer's actions is recorded in the database entity Customer–order.

There is no guarantee that data analysis methods will produce entities which map directly onto JSD entities, but both approaches do emphasize modelling the structures in a system, either static or dynamic, rather than surface functions. JSD may therefore help analysis of database structures.

Actions constitute entities and are the fundamental building blocks of the JSD model. JSD actions may be considered to be synonymous with events, although they are a hybrid concept with action in the more literal sense of activity. Actions describe how, and to an extent when, things happen, and to do so they need to be bounded by stop and start events to determine their sequence and timing. An event is the point in time when something happens. Hence a start event for boiling an egg could be putting it into boiling water and the finish event is taking the egg out of the water. In between these two points in time the action 'boil' is being done to the egg. In JSD, actions have a very short duration, and hence the distinction between action and event is not made.

When analysing actions, stop and start events may be found to be close together in time or separated by an indefinite period. JSD prefers actions to be short and precise, so if an action has indefinite stop and start events it is a poor specification and should be re-examined to see if there are other events which subdivide it more clearly.

The scale of time used to determine actions will depend on the application. In information systems actions may take place in an amount of time which people can perceive (e.g. placing an order); but in real time systems actions will be more atomistic and closer to instantaneous events, e.g. a switch changing in microseconds.

Actions cause change in a system, and specification of when change happens in a system is important in JSD. If we model the 'what happens when' and include all the possibilities of what could happen in a system, then the programs which implement the system should be able to handle all possible processing sequences without error. This emphasis on time ordering in JSD guides the analyst to produce a correct design.

Analysis of change over time is recorded as sequences of actions in process structure diagrams (PSDs). The details of PSDs are described in the next section; for the present the important concept to note is that actions are shown in a time-ordered sequence in PSDs. How the sequences are ordered is determined by dependencies between events which help the analyst to place actions in a time order. Often this is a matter of common sense; for example, in sales order processing Check(customer credit) should not happen before Find(customer ID on reference list). The action to validate that the customer is on the company customer file precedes the credit check action.

Ordering actions will result from fact-gathering during systems analysis; however, the JSD interest in time ordering makes the analyst approach such activity with different questions from those employed with traditional methods. In particular the analyst needs to ask questions about 'In which order is X or Y done?' and uncertainty 'Is X always done?' or 'Is it X sometimes but Y normally?'. JSD actions are also at a quite detailed level of specification, so several interview sessions may be required before the necessary facts have been gathered for precise specification. Each JSD action should be a *primitive*, that is, a non-decomposable unit which does one thing and has no events of interest contained within it. This definition may appear a little vague at first but experience soon gives the analyst a better feel for the concept.

Actions will eventually become the building blocks of programs; if they are correctly analysed then the behaviour of the program should be correct. The program should also be relatively easy to modify because the location for change is made explicit in its structure. If actions are incompletely decomposed or inaccurately specified, then location of a modification (for instance, a calculation which has to be changed) may be difficult because it is buried inside another component. This often produces undesirable side effects as change to one part of the system upsets another hitherto correct part. Identification of actions and entities is central to JSD modelling of a system's activity.

2.3 PROCESS STRUCTURE DIAGRAMS

Process structure diagrams (PSDs) are tree hierarchy diagrams which read from left to right. The trees are composed of leaves called *elementary components* at the bottom of each branch and *group components* higher up the branches which meet at the top of the tree. Only actions can be leaves; the rest of the tree is used to group actions together and to obey diagram conventions (see Figure 2.2). In such diagrams the convention is to call components at a lower level the *children* components or *sons*, and the component immediately above the children the *father* or *parent*; all the components belonging to one father constitute a *family*. Hence all group components must be parents, while elementary components cannot be parents.

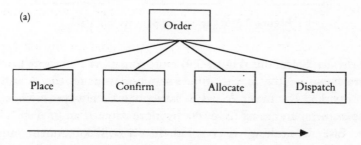

The order is first placed and then confirmed, goods are allocated to the order and finally it is dispatched.

Figure 2.2(a) Process structure diagram: sequence of execution of actions in the Order process.

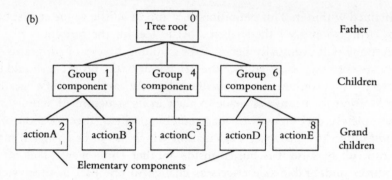

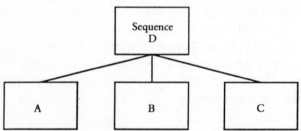

Figure 2.2 (b) Process structure diagram: composition. The reading order is shown by the numbers in the top right-hand corner of the boxes.

The reading sequence goes up and down subtree branches progressing across the diagram. There are four component types in PSDs.

(a) *Sequence*. A sequence may be composed of one or more parts which follow each other from left to right; in Figure 2.3 A is followed by B which in turn is followed by C. Note the hierarchical meaning in the diagram: the upper component D is composed of the sequence ABC.

Figure 2.3 Sequence components in a PSD.

(b) *Selection*. Selections (Figure 2.4) must contain two or more parts and are exclusive; that is, each time a selection is made *either* A *or* B *or* C is invoked but never A *and* B. Selections are always 'If Then Else' constructs, and must have two or more parts. If an 'If A = B Then X Else do nothing' is required then a null component must be added. Note that the conditional logic will be contained within the parent component D and that the actions selected are marked with a circle in the box. This apparent contradiction becomes clearer when the diagrams are compared with the equivalent text notation.

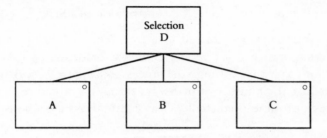

Figure 2.4 Selection components in a PSD.

(c) *Iteration*. Iterations consist of one part only and the repeated component is marked with a star in the box (see Figure 2.5). As with selections, the conditional logic (do while etc.) is in the parent component D. Iterations may occur zero or more times and this can be used to model the possibility of an event occurring (zero times means it does not occur).

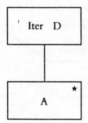

Figure 2.5 Iteration component in a PSD.

(d) *Null*. This null or do nothing component is represented by a hyphen in the box (see Figure 2.6). It is used in selections when one part is a 'do nothing' option. This is used to model the possibility of an event not happening, in which case the null side of the selection is invoked.

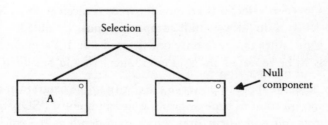

Figure 2.6 Null component in a PSD.

Four commandments govern the construction of PSDs.

(a) No cross-links are allowed in a tree. This would upset the time ordering, which reads up and down the branches, by creating two or more pathways through the process (see Figure 2.7). As JSD aims to reduce the number of pathways through a structure to make its behaviour more predictable, clearly cross-links are a nonsense.

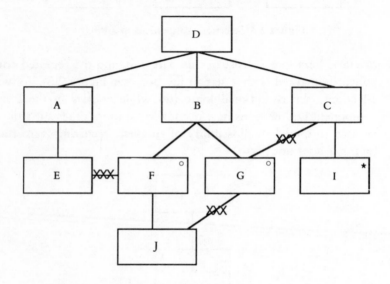

Figure 2.7 PSD tree diagram showing illegal connections (marked X).

(b) Within one family the sons must be of the same type (see Figure 2.8). This is because to mix a selection or an iteration in the middle of a sequence is logically impossible. One decision has to be taken before the next action can proceed. The solution to the problem of mixing selections and sequences is achieved by introducing an extra component into the diagram.

(c) One-part selections are not allowed. If faced with an 'If X then do A' with no 'else', a null component must be added.

(d) One father can have only one iterated son. If a sequence or selection is to be repeated an extra component should be added.

The rules are there to preserve the clarity of the diagram and ensure that the specification of time ordering is unambiguous. PSDs start out being shallow and wide structures which gain depth as the rules are applied and higher-level components added to group actions together. The

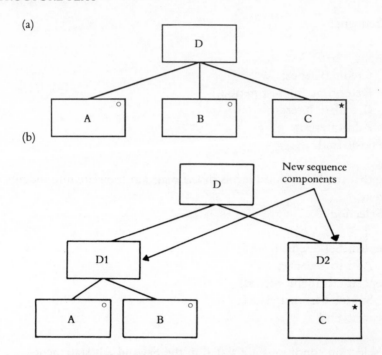

Figure 2.8 (a) Illegal PSD tree with mixed sons; (b) correct version of tree – an extra layer of sequence components is added to prevent one father having mixed sons.

resulting diagram is clear to read and excellent for walkthroughs to verify a specification.

2.4 STRUCTURE TEXT

Structure text is a text form of PSDs. Its purpose is to produce an alternative form of specification. JSD specifications can be either diagram-based, in which case extra detail is annotated on to diagrams and cross-referenced to lists, or text-based.

The text specification mirrors the component structure and hierarchy of PSDs by using indented paragraphs. In JSD structure text may run in parallel with PSDs as the specification progresses or may form the main means of representing the specification. The choice of representation method is left to the analyst.

The PSD components are translated into text form as follows.

(a) Sequence

    Deposit Seq
        Credit Balance;
        Determine Interest period;
        Calculate Interest;
        Add Interest;
    Deposit End

Note that the hierarchical levels of the diagram translate into indents in
the text.
(b) Selection

    Deposit Sel (credit record)
        Add to Balance;
    Deposit Alt (debit record)
        Subtract from Balance;
    Deposit End

The selection conditions are stated in the Sel and Alt statements.
(c) Iteration

    Deposit Iter (while credit records to be processed)
        Add to Balance;
    Deposit End

The condition for iteration is stated. Conditions are always 'do while' or
'test before', and the iteration may be invoked zero or more times.
Sometimes it may be necessary to use the condition Iter (while true) to
specify iterations which may continue for ever. Actions, which are
elementary components, are terminated with a semicolon.

An important difference to note when mapping PSDs on to structure
text is that the diagrams show the Selected or Iterated component
marked as a Selection or Iteration, whereas the text has the Selecting or
Iterating component marked with the component type name of Sel or
Iter. This may appear to cause problems when Selections have sons
which are Iterations and vice versa, as in the structure shown in Figure
2.9.

The solution is to add additional components – Account Body and
Update Body – to the diagram and structure text:

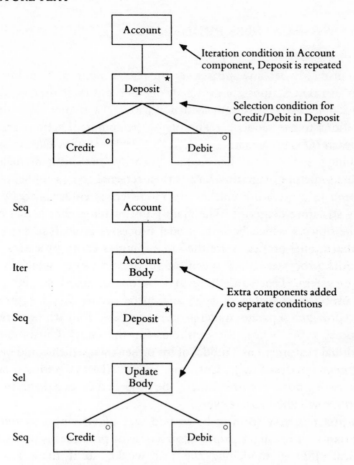

Figure 2.9 PSD with selections and iterations.

```
Account Seq
    Read(Transaction);
    Account Body Iter (while Transactions)
        Deposit Seq
            Update Body Sel(credit Record)
                Credit;
            Update Body Alt(debit record)
                Debit;
            Update Body End
            Read(Transaction);
        Deposit End
    Account Body End
Account End
```

2.5 LONG-RUNNING PROCESSES

It has probably become apparent to the reader that time is an important factor in JSD. Actions are grouped together and their time ordering is recorded in a tree diagram which illustrates a sequential order and exceptions to the sequence. The major groupings of actions are called entities in JSD. Entities have a Jekyll and Hyde identity. During analysis the thing which is first identified is the entity, and this is modelled in a process structure diagram which is then referred to as a model process. In some JSD literature entities are modelled as entity processes with entity structure diagrams. The main point to remember is that entities are the objects which become model processes as analysis progresses, and that model processes are the same thing as entity processes.

Model processes are a specification of something which happens in the system. That something may be one or many somethings, so JSD uses the concepts of type and instances to describe processes. Conceptually a separate model process is created for each occurrence of a process type; hence, for example, for the entity Customer every individual customer has a model of his or her own which could be called Customer process(Fred), Customer process(Mary) and so on. All occurrences, however, are going to be modelled on one diagram which will represent the process type.

At first this may seem to be an odd way of viewing a system. The importance of having separate occurrences of processes is partly philosophical. JSD is modelling the real world, so if there are many occurrences of one type of thing then the model should reflect this. But there is an important practical benefit. JSD causes us to analyse how actions happen over time, and by considering multiple occurrences we have to analyse not only entities with normal life histories but also ones in which exceptions and unusual events happen. Good analysts have always recorded not only the norm but also the exception conditions in systems, but in JSD such questions are raised explicitly by the method.

In the JSD specification stage, individual occurrences of processes conceptually may or may not exist. Consider a bank account example; when a customer, John, walks into the bank and asks for a bank account, a new instance of the Bank-Account process is created for John and its first action is invoked (Open). John then uses his account and many debits and credits are performed on his account. This phase of the Bank-Account(John)'s life history is a repetition of Credit and Debit actions. John finally gets fed up with the bank and takes his business elsewhere, invoking the last action in the life history – Close. The occurrence of Bank-Account(John) is then deleted from the system. Individual processes may be continually created and deleted during the lifetime of a system and the model has to account for that fact. Sometimes individual processes are kept even though their life history may appear to have ended; in this case they form an archive or history about the system.

The time ordering of actions within processes is shown in process structure diagrams which explicitly represent the order in which actions occur. There is usually only one diagram for each entity (although entity roles may create more, as described in Chapter 3); hence it has to show all stages in the life histories of all occurrences of the entity type.

The concept of many occurrences of one process type is maintained throughout the JSD specification until implementation, when a single process which models all occurrences is physically designed. This transformation is simple; all that has to be added is a record (in the computer and the everyday senses of the word) for each occurrence of the process. The record contains the history of what has happened to each occurrence (e.g. Account(Mary) is in the black, Account(Fred) is in the red). The physical account process has to be able to deal with all the possible actions which can happen to Fred's and Mary's accounts.

JSD processes are viewed as being 'long-running'. This is to draw a distinction between modelling something that has a complete life history and a procedure that models a function, i.e. something which carries out a specific task in an identifiable time period. JSD is interested in modelling everything that can happen to the entity, and therefore it has to deal with events now, tomorrow and at all points in the future. Processes have to exist as long as the modelled entity exists and in this sense they are considered to be long-running.

That does not mean that JSD processes are in a state of perpetual activity; most of the time they will be in a wait state anticipating arrival of the next piece of data to process. A process will have several actions during its lifetime. Actions will either be triggered by the arrival of data or trigger each other in sequence until the final action is complete, whereupon the process will notionally terminate and cease to exist. If in the banking example Mary closes her account, the process Account (Mary) will cease to exist. In the physical design the process type will delete Mary's account record and return to its starting action to deal with the next account that becomes active.

2.6 SUMMARY

Central to the JSD approach is the concept of modelling systems. The model focuses attention on system structures rather than on surface manifestations, usually identified as functions. The system model is based upon a series of concurrent communicating processes. Each model process, initially identified as an entity, is composed of actions which are ordered into a time-ordered sequence. Actions are the atomistic building blocks of the specification, representing operations which cause change to data, and cannot be subdivided further.

Each model process is viewed as a type which may have many occurrences. It is an objective of modelling to record the life histories of all occurrences of a process type. In this way normal and exceptional life histories are recorded. JSD processes are conceptually long-running; this is different from procedures that are a set of task-related activities which take place in a definite period of time.

JSD makes extensive use of diagrams. Process structure diagrams are used to model entity processes and show their constituent actions ordered in time. Time ordering is expressed using four basic action types: sequence, selection, iteration and null. Structure text mirrors the PSD and is an alternative form of documentation.

Key points

- JSD models objects within systems, called entities.
- Entities respond to events in the real world.
- Entities are composed of actions, essentially responses to events in the real world.
- Entities are modelled as a process type and may have many occurrences of the type.
- The model process of an entity is considered to be long-running, i.e. it has a long life history.
- Actions within entities are modelled in time order.
- Time ordering is expressed in process structure diagrams.

3

JSD MODELLING STAGE
Entity analysis

This chapter deals with the modelling stage in JSD. This is the first phase of JSD, and it identifies and describes the major structures in the system. Heuristics for identifying entities and actions are described and the qualities of JSD entities are reviewed.

3.1 DESCRIBING ENTITIES

Discovering entities within a system is one of the more difficult tasks in JSD, and one which requires considerable experience. It is also possibly the step in JSD where inexperienced analysts are most in need of help. Before tackling the question of entity analysis it is necessary to have a clear view of how JSD defines and models entities.

JSD entities are composed of actions. Actions are discrete nondecomposable steps which form the basic building blocks of the system. An action describes what an entity does within a system, whereas the entity is the 'thing' that a set of actions have in common and can be thought of as a task or group of actions within the system. Another view, with elements in common with object-oriented programming and abstract data types, regards an entity as an object of interest in the world which has a set of data descriptions and actions associated with it.

Entities can model people's roles in the real world or, equally, inanimate objects such as switches and sensors. Entities may be models of active things that the system has to respond to; on the other hand, they may be objects that the system will change. A wide variety of objects can therefore be entities and the concept is probably best explained by example. For instance, some people-type entities could be Sales Clerk, Customer or Invoice Clerk. Inanimate entities could be mechanisms that also do things, such as Servo Control or Autopilot. All have in common actions that alter objects (usually data items) and influence activity in other parts of the system. Thus the Customer places, amends or cancels an order; the Autopilot increases the speed, turns the rudder or raises the elevators. Note that all these actions are expressed in the active voice of the verb.

Other examples of entities could be more conceptual, such as Order, Invoice or Bank Loan. These represent documents, and transactions model the life history of these objects as they interact with the system. Entities may either perform or suffer actions. This means that an action may be performed by an entity which will influence something else in the system, or that an action may be suffered by an entity in response to an influence. Whereas performing actions are recorded in the active voice, actions which entities suffer should be in the passive voice because something is being done to the entity. Hence the Order, for instance, is 'placed, amended or cancelled'.

Entities do not always follow a single life history of actions, however, and in some cases it is possible to take one or more views of the same object. The action performed and suffered by the object depends on the context. In this case we need to model entity roles.

Entity roles

Entities may occur in different roles. In some cases these roles will be related to a classification hierarchy, while in others they will depend on the viewpoint or context of what is being described. To illustrate: vehicles in a road tax system may play one role, but within 'vehicle' it may be possible to identify other roles for goods vehicles, which have extra actions associated with HGV licensing, and also within 'goods vehicles'. This hierarchical approach could be taken further with vans and lorries having subroles.

Entities like these were described in earlier versions of JSD as *generic entities*. This is a super/subentity concept in which the subentity is a member of a larger whole even though both have a separate series of actions and distinct life histories. For instance, subassembly could be part of Automobile, and Part could be a member of Component. It is better to focus on the concept of roles rather than on hierarchy because JSD wishes to define the activity of objects rather than simply to classify them.

In some cases one object may have two or more roles although the tangible thing is the same. For example, an invoice might be used both for interdepartment billing within a company and for external billing; the two roles will have separate life histories even though the same thing is used for both roles. Consider, as a further example, the multiple roles of a person in a marketing survey system: one role might describe a person's career promotion and earning potential, while another separate role could model how the income is disposed of.

Finding entities

Discovering entities is difficult, so we need to consider some basic JSD principles to guide us. First, JSD is about problems, modelling and the specification thereof. The first questions we should ask are:

- What problem is this setting out to solve?
- What is the goal this system aims to achieve?
- What are the objects the system responds to or manipulates?

JSD encourages us to model our system on the real world, so a good starting point is to look for real world objects with which the system will have to interact. The answers should lead us to identify one or more entities which will describe the central problem. Thus a library system is all about Books, the lending and return thereof; a lift system is all about Elevator Controls, the calling and the scheduling involved with it; an order-processing system is about Orders, accepting them, handling cancellations and so forth. The central entity is going to describe the life history of the main thing which the system exists for, and the description states the problem.

Another approach is to use actions, listing things which happen in the system as simple statements, phrases and sentences. Without conscious intent on the analyst's part, many phrases will refer to things within the system which can be identified as entities. Hence in the library example the statements 'members borrow books' and 'members must have a registration card' points to an entity Member. In order processing the statements 'parts are allocated to orders' and 'damaged parts are returned to stock' suggest an entity status for Part and possibly for Stock. The approach is to list the phrases, identify phrases with nouns in common, and then examine the nouns for the objects they identify as potential entities.

One problem with phrase listing is that synonyms may obscure the identity of an entity. For example, in the order-processing example Part, Component and Stock may be a collection of nouns related to stock movements. Related nouns need to be examined to decide if they are describing one entity or, as is frequently the case, the different labels are describing alternative roles for the same entity.

Another heuristic which may be useful for identifying entities is to ask the question: 'What are the principal actors that help to solve the problem?' The answer to this question should identify other entities which are necessary to help solve the problem. This question helps identify the 'people-type' entities which interact with the system. Once again the best illustration is by example: the library system may need a Member entity who borrows and (it is to be hoped) returns books, while in an order-processing system a Clerk may be necessary to allocate orders for delivery.

Another hint for discovering entities is to look for all the jobs or tasks people perform in the current system and the documentary evidence for their work, i.e. the transactions they process. Alternatively, if the system is inanimate, look for the mechanisms inherent in it and the operations that must be performed to operate the system. It is important to remember that an entity is essentially an organized activity, composed of a sequence of actions grouped together by a common inter-relationship. The strategy of looking for entities in jobs and tasks is something of a double-edged sword and the analyst should be beware of making analysis too 'people-centric'. Two people who share a job or a logically related job may be split up for management reasons. The analyst has to be constantly aware that objects are being modelled, not short-term tasks; otherwise functions and not entities will be described.

Entities and functions

The difference between entity and functional modelling is important in JSD and deserves further attention. If something is organized there is a good chance it has been recorded – perhaps as a job description, procedure or operating instructions. Procedures, work plans and instructions will be needed in the JSD design, but including them now could create an error. These phenomena are likely to be functions (see Chapter 4) and are recognized as being a group of actions which execute in a relatively short space of time in order to achieve a specific task. Such functions should be set aside for inclusion at a later stage of design. An experienced JSD analyst will immediately recognize functions and not model them at this stage. For the inexperienced the best advice is probably to remember that the entities to be modelled will be long-running processes, i.e. they have significantly long life histories. To use another metaphor, one needs at this point to model only the actors who are going to appear on the stage throughout the drama.

Procedures, in contrast, have short life histories because they are part of someone's job and can usually be completed in a short period of time or at least before the end of the day. An approach to differentiating entities from functions is first to model things which change, or cause change, throughout a significant part of the system. These will be entities. Anything which is a discrete procedure or an algorithmic set of operations should be set aside as an interactive function and returned to later in the design. Examples in a library system could be procedures for cataloguing books and for eliminating old books (after x years disuse, say), although note that the action recording the effect of the procedure Delete would be recorded in the entity Book. In order processing the stock allocation algorithm (e.g. 'if not in stock, then fulfil part order') is a good candidate for the role of an interactive function.

A definition of the difference between entities and functions is that entities represent groups of time-ordered actions which describe a significantly long life history of something within the system, whereas functions are a set of actions which take place in a short space of time to accomplish a task. This definition may become clearer with a metaphor to illustrate it. Imagine the system is a fish pond: JSD entities are the objects of interest deep within it, i.e. the fish, while functions appear on the surface as manifestations of the existence of those deeper structures, e.g. the ripples on the pond when a fish rises.

Surface functions are tasks which can usually be easily identified. Behind these tasks there is a 'thing' which describes that part of the business's or organization's life upon which the task or procedure may have an influence. For instance, in sales order processing the order is changed into a discounted sales order by a discount calculation algorithm. Using the pond analogy, a heuristic to consider when approaching entity analysis is 'try to look beyond the surface on which functions are visible to see the entities below'.

To give a further example: an order-processing system may be described to the analyst as functions for Order entry, Order allocation, Order dispatch, Stock movements, and Stock replenishment. These functions are the surface representations of events which happen to JSD entities Customer, Order and Stock. It is the entities' job to describe the life histories of things which are important to the business. JSD maintains that correct models of systems activity can only be derived by considering the underlying structures in systems, and that analysis of the surface functions merely records the present solution to a problem.

Steps in entity analysis

The initial analysis will probably produce a list with too many entities. The next step is first to decide which entities are inside the system and therefore worthy of further investigation, and then to eliminate any duplicate or phantom entities ('phantom' entities are entities which really neither perform nor suffer actions and are therefore just static data).

Another problem arising in entity analysis is the boundary problem. Where does one entity finish and another begin? This problem should resolve itself when the actions for each entity are time-ordered (see Section 3.3). Phrase lists also help with the boundary problem. The matching of actions to entities should discourage any overlap. In phrases the entities will appear as subjects and objects of the verb; for example, 'Clerks allocate Parts to Orders' suggests three entities.

The chances are that the entities identified in initial analysis will require considerable modification. To decide if the entities identified in the first pass are satisfactory or even necessary, we need to allocate actions to entities and describe their time ordering.

3.2 DESCRIBING ACTIONS

JSD actions happen at a point in time and are atomistic, i.e. they cannot be decomposed further and are uniquely identified. Actions should occur in a short period of time – just how short is not explicitly defined in JSD, but basically the shorter the better. So the action Dispatch starts when the clerk picks the delivery note up and puts it in an envelope, and finishes when it is placed in the out-tray. To illustrate the point further: Place, Cancel and Amend (an order) are all fairly short actions which take place at a point in time, and each may be considered as a single event which can happen to an order.

In contrast an action which is continuous or ambiguous is a poor description. For example, processing, sleeping, absent are states of which the start and stop events are ill-defined and separated by a considerable time period. It is advisable to think of concise short actions which can be described as events in the life history of an entity. Another point to bear in mind when vetting potential actions is to reject output actions such as Produce report; these are likely to be part of the functional processing within the system. Actions respond to input messages.

Actions are drawn up in lists which describe each action with a sentence or two of English and list the data which are necessary for the action. Actions respond to events in the real world. These events are communicated to the system as data messages and are referred to in JSD as the *attributes* of the action. Hence to discover actions the following heuristic may be used:

(a) Decide on the external events which happen in the real world.
(b) Define how these events are communicated to the system as input.
(c) Identify the system inputs (these are action attributes).
(d) Define actions which are necessary to respond to the messages contained within the action attributes.

Actions are listed with their associated attributes. The actions can then be associated with entities.

Action list
Place: customer places order with sales clerk by phone, letter or in person.
Dispatch: warehouseman dispatches goods and delivery note to customer.
Return: customer may return damaged goods to the warehouse.
Bill: accounts send an invoice to the customer.
Amend: customer may change the order details.

Entity action list
Entity ID: Customer

Actions

Place: customer places order with sales clerk over the phone, by letter or in person.

(Attributes: customer name, address, product ID, quantity, date, place status.)

Amend: customer changes order details as above.

(Attributes: order number, new order details, amend status.)

Cancel: customer cancels an order. This can only be accepted within 14 days of initial placement.

(Attributes: order number, date of order, cancel status.)

Most actions will belong to one entity but some 'common' actions may occur in more than one entity. In a library system 'Borrow' may be an action of the Book and Library-member entities. Common actions will create a link between the entities that share them. This property has two important consequences: first, it creates a relationship between the entities (see Chapter 5) and, secondly it expresses a constraint. Hence the Book entity cannot execute a Borrow action before the member has borrowed the book.

It may be necessary to resolve conflicts about which actions belong to which entities. Sometimes this will be clear but some actions may seem to be suitable for two or more entities. To answer this question we need to determine the entity boundaries. These will be the first and last actions in a life history, so to draw the entity boundaries we need to order actions in a time sequence – for example, Place order, Change order, Confirm order, Dispatch order.

To start with, actions should be ordered in a list in rough sequence of their occurrence; the exact ordering in the middle is not too important at this stage but the first and last actions must be identified, because these determine the entity boundaries. Most entities have a discrete sequence with beginning and terminating actions and those that do not often have a simple structure. For example, the Button entity in a lift system could be a simple iteration of presses and releases. To check actions and entity boundaries the following heuristics may be used:

(a) Order actions within each entity in time and attempt to find a start and finish action for each entity.

(b) Ensure all actions can and should be logically carried out by that entity.

Action lists contain an English language description of what each action does. This is a useful check, because if an action cannot be described in a sentence or two then its existence is debatable.

Marsupial entities

Sometimes entities may have sublife histories within them which have to be separated out. The problem is essentially one of concurrency with different actions going on at the same time within one entity. Take the example of a bank account. If one person has only one account all is well. We can model, say, Account(John) and be sure the actions Open, Credit, Debit and Close will occur in a certain order. But if one person can have two accounts then we cannot say whether a Credit is for Account(John-1) or Account(John-2). We may also have two events to process, a Credit(account-1) and Debit(account-2), in the same model process. The solution is to remove the subentity which has a separate life history; in JSD this is called a *marsupial entity*.

When reviewing an entity for possible marsupials a useful question to ask is: 'Does an instance of this entity remain as one instance throughout its life or can it give rise to one or more separate children?' An example is the Order entity, which gets split up into part-orders when there isn't enough stock. In this case we have a new (or marsupial, in JSD) entity Part-order, which we need to split off from its parent. The marsupial entity will share common actions with its parent. A point to note is that marsupial entities are often repeating groups of actions within an entity, which are separated in a similar way as repeating groups of data items are separated from entities in first normal form analysis. Discovery of marsupials automatically does a first normal form analysis on the entity attributes, a point of data analysis which is returned to in Chapter 5.

Entity attributes

Early versions of JSD had little advice to give about data analysis, and attributes were merely listed; however, more recently data have received more attention.

JSD distinguishes between *action attributes*, which are input data messages consumed by actions, and *entity attributes*, which are descriptions of an entity. Entity attributes are created and updated by an entity's actions and as a result record the entity's life history. Entity attributes are internal variables and can be thought of as a kind of status record of an entity, i.e. the information an entity has to store so that it knows where it is within its own life history.

An entity attribute should have a meaningful name and a type. A type describes the form the attribute should take. In JSD the meaning of *type* is slightly different from the usual programming idea of type, e.g. integer, real or character. The attribute type is used to refer to the origin of the attribute. Thus in a library system with two entities, Book and Member, within the Book entity attributes Borrower-ID records who has taken the book out of the library. Borrower-ID would be given the type Member because it is derived from that entity. This concept of type will prove useful in data analysis for identifying foreign keys within entity attributes.

The update conditions of entity attributes are specified by defining the actions which can update them; as a result the entity attribute specification becomes a precursor of a database structure with update constraints. One entity attribute is required as an identifier for the model process; this will be the key field and cannot be updated by the entity itself. All other attributes are updatable by the model process which owns them according to the constraints specified by the entity's actions.

3.3 TIME ORDERING ACTIONS

A rough estimate of time ordering has been carried out to decide on entity boundaries. This step takes the approximate ordering and analyses it further to create an accurate picture of what happens during the lifetime of an entity. In doing so the entity is translated into an entity structure which describes its life history. These entity structures become model processes.

Time ordering is bound up with constructing process structure diagrams which illustrate the relationship of actions within a process. Actions are shown as the leaves (i.e. the lower-level components on the diagram); components higher up represent higher-order aggregations of actions which are created partly out of choice and partly to obey diagram rules (see Figures 2.2 and 2.8). While making PSDs some actions may need to be analysed in more detail, either because they are too vague or because they encompass too much activity. On the other hand, some actions may appear too trivial and may be eliminated by merging with other actions.

Time ordering is an iterative process during which mistakes and insights are revealed as analysis proceeds. The analyst should expect time ordering and process structure diagrams to go through several drafts until an accurate picture is gained.

Actions are defined in JSD as being atomistic (that is, they cannot be further subdivided), but for inexperienced analysts the point at which to stop subdividing may be far from clear. There is no guaranteed solution to the problem of 'Where do I stop splitting actions?', but a rough guide is that an action should be expressible in a line or two of English and should not possess any significant time ordering (iteration or selection structures) within it.

To illustrate the point: an action such as Create(sales order) may have several substeps within it, such as fill in customer details, enter date, write out products ordered, quantity and cost. Such a sequence may be permissible as a single action. But if the action has significant time ordering within it, in particular choices or repetitions, then it should be decomposed. So if within Create(sales order) many products were to be ordered and a quantity discount calculated, then new actions would be required – Enter details, Calculate discount.

The content of an action will also depend on the level of resolution of time needed for the application. In real-time systems a very fine event division may be needed, but in information systems such fine resolution will be unnecessary. A rule to bear in mind is that JSD systems are about recording events of significance. If, having examined the set of events the entity has to respond to, it could be reasonably expected that the events could not be subdivided, then the action is correct in JSD terms. On the other hand, if the action does not provide the system with enough information about the real world object then the action needs decomposing.

To start time ordering, a top-level box or diagram node of the process structure diagram is selected. Actions are then ordered using the facts recorded in the action list. JSD has no specific procedures to follow when time ordering. The following list of heuristics may be used as an approximate guide:

(a) Start by ordering actions in a sequence from the first to occur in the entity's life on the left-hand side of the diagram to the last action placed on the right-hand side.

(b) Add any choices and repetitions to the diagram, and reorder to obey diagramming conventions. This will add selections and iterations to the PSD and will naturally add more layers or depth to it as the linear sequence is changed.

(c) Add any high-level components which may be needed to make the structure more comprehensible. This may not be necessary; in a PSD which consists of a flat sequence of actions, however, grouping will help to make the diagram easier to read by adding hierarchical structure, e.g. 'Opening actions – Middle sequence – Closing actions'.

(d) Criticize each action definition to ensure that it is neither too trivial nor too large. If actions are unsatisfactory either subdivide them or eliminate them from the PSD.

3.4 DEALING WITH ALTERNATIVES

JSD places considerable emphasis on designing processes which can react to many different events in the system's environment. Design of alternative processing pathways in PSDs therefore deserves some attention. Choices in a process life history may be found as an explicit statement during analysis, perhaps in the form of a written job procedure such as 'Either do A or do B'. More frequently choices are implicit, and further analysis is required to discover alternative pathways. A good example of this is a career structure which could be found in a personnel system.

A first pass structure might look like Figure 3.1. Instances of the career entity in this case are individual employees whose careers could be at one of several alternative grades. A multiple-part Select is used because the person can become either a programmer, an analyst, a project leader and so on.

The first career PSD structure, although it may work for a limited set of conditions, does not contain much information. There is no way of knowing from this model whether someone's career progressed from an analyst programmer to analyst or from a systems programmer to analyst. The specification is a poor description of what really happens to a career history. In addition the process could be prone to errors such as a trainee starting as a project-leader, something which is against company rules.

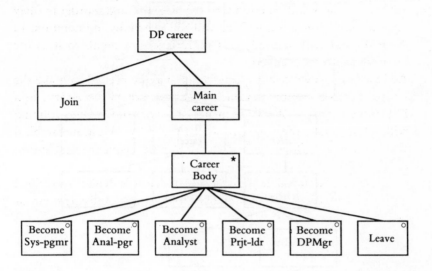

Figure 3.1 DP career process: first PSD. The flat structure gives little information about the entities' history.

A better description is shown in Figure 3.2. Here the multipart selection has been refined to give a sequence of selections which model the career history. A person's hiring as a trainee programmer becomes the first event in the career. All trainee programmers then follow one of three routes: they may either progress to being a programmer, leave the company voluntarily or be fired (assuming no one remains a junior programmer for life). The PSD tree is reorganized to show each career step as a nested selection. This describes the career order more accurately (i.e. no one goes from a trainee programmer to project leader overnight) and preserves the choice at each step. Note that after the trainee programmer the selection is 'progress to next step/leave/get fired/ retire', so a person could remain a systems programmer for a long time until leaving. The PSD says nothing either about the time between steps or about the most probable career path, even though it accurately models all the possible career paths.

At each career step the diagram has to model all the possibilities so that there are exit points from the structure. If this is not done it would be possible to get trapped inside the structure and never traverse the PSD tree, which would be a model of a never-ending life history. Figure 3.2 is therefore wrong because we may have a junior programmer who joins and becomes a systems programmer, remaining that way for life.

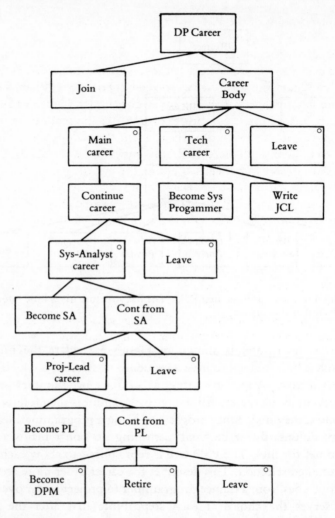

Figure 3.2 DP career process: second PSD. This structure is properly modelled, using backtracking as explained in Section 3.5 and shown in Figure 3.5.

When the day comes to pension him or her off we have no action to deal with that event. The PSD has to be able to deal with all the possibilities which could happen to the life histories of all the individuals it is supposed to represent, even individuals as unlikely as the systems programmer with no ambition.

The analyst has constantly to ask the question 'What could happen at this step?' to tease out all the possibilities, and a good way to verify these steps is to ask the question 'Can the life history get stuck at this step?'. If it can, elaboration of the PSD is required to provide an escape route so the life history can come to an orderly end.

PSDs are very useful tools for analysis walkthroughs to ensure process descriptions are correct because they are WYSIWYG diagrams (literally 'What you see is what you get'). For example, in the career structure what happens if an employee goes under a bus? No allowance has been made for this career event – a premature termination of the career due to an unforeseen accident. Premature termination is a problem in many life histories. An entity may follow the normal sequence of events but at certain points the unexpected may happen. In the DP career entity an individual may leave at any stage because he or she gets another job, and there is no way of predicting when this may happen. This problem has a special solution in JSD called 'Backtracking' which is dealt with in Section 3.5.

PSDs can be checked against input data to ensure that all the messages coming in can be processed and conversely that all the possible events which the model process has to recognize are recorded within input messages. This is an exercise where there is no substitute for experience, but a critical reading of PSDs and thinking out all the eventualities helps to clarify the analyst's, and the user's, understanding.

Sometimes an alternative may have to be expressed that is not of the 'either/or' type; this alternative may be an event possibility – something may or may not happen. Because of diagramming conventions, selections have to have two parts, so one component is marked as a null (or continue) to take care of the possibility that the event does not happen and the other part is invoked when the event does happen (see Figure 3.3).

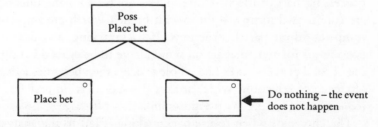

Figure 3.3 Dealing with possibilities: use of a null component in a PSD.

Once alternatives have been examined the PSD should be reasonably accurate. It should show how the entity changes throughout its life and map all the possible pathways the life history can take, including events which are unlikely.

3.5 DEALING WITH THE UNEXPECTED

The unexpected is an extreme at one end of a continuum with possible alternatives in process structure specifications. Not only do we have to deal with all the fates we may expect for an entity; we also have to plan for the unexpected, the unusual or when things just go wrong. If the unexpected becomes a dominant feature of a process and there is possible premature termination of the life history then JSD uses a special technique to deal with uncertainty called *backtracking*.

The problem is that we cannot determine when a process will come to an unexpected end until we get the necessary information in the input data. This uncertainty at runtime could be modelled by selections for every action with input, one part dealing with the normal event, the other part with the unexpected event. Although possible, this modelling leads to an untidy process structure. Backtracking is used to reform untidy nested selections which contain multiple tests for uncertain conditions and for iterations in which completion of the iterated actions is not certain.

Backtracking creates a process structure which has two parts, one of which deals with a normal sequence of events but also tests for the unexpected, and the second part which takes any necessary remedial action when the unexpected has happened. The JSD (and JSP) terms are Posit (all is going well in a normal sequence), Quit (if something has happened) and Admit (something bad has happened and remedial action is taken), as illustrated in Figures 3.4 and 3.5.

Quit statements are embedded at strategic points in the normal processing part of the structure. These are tests for the unexpected and are conditional jumps to the Admit branch which are implemented as Jump-to-Admit part instructions. If backtracking was not used small procedures for remedial action would have to be embedded throughout the normal process text. The whole process then becomes a deep-nested sequence of selections which makes the specification text less tractable (compare Figure 3.2 with Figure 3.5).

The choice of when to use backtracking is left to the analyst. Minor uncertainties can be dealt with by an ordinary selection and a null component if necessary; but if there is considerable uncertainty in a structure and the outcome of tasks can only be determined at run time, then backtracking is advisable. Uncertainty at run time is often caused by qualities of the input data, because there is no way of predicting whether it will be correct or not. There are many situations when this

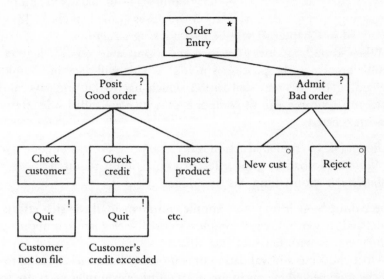

Figure 3.4 PSD showing backtracking. ! indicates Quit; ? Posit/Admit; + and −
can be used to indicate Beneficial/Neutral and Intolerable side effects respectively.

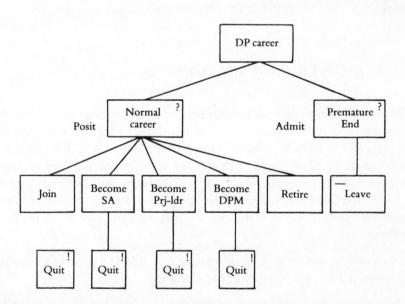

Figure 3.5 Backtracking structure for DP career: the normal career is modelled as a
sequence with Quits to detect premature termination of the career. The Quits and
the Admit branch (Premature End) could be made more elaborate to model different
premature ends (leaving for another job, sickness, etc.).

may occur but one of the most familiar is in data validation; not surprisingly, most data validation processes (called filters in JSD and described in Chapter 4) will be backtracking structures.

When something does go wrong and a Quit statement is triggered the Admit branch of the process is invoked. This will vary in complexity depending on the remedial action which has to be undertaken. JSD distinguishes three Admit components, which are called side effects (of the unexpected):

- *Beneficial*. No remedial action is necessary, the error is in your favour.
- *Neutral*. No bad or good effects so no action is required.
- *Intolerable*. Something has to be done about it.

The Admit branch may be a simple comment if all the side effects are beneficial or neutral, or a complex subtree with many components to handle errors with harmful side effects.

Most checking and validation operations will use backtracking to test for the unexpected events in input data. Other examples of its use could be process control emergency procedures in a monitoring system and a monitor with recovery/restart procedures in a computer operating system.

3.6 ELEMENTARY OPERATIONS

Detailed design in JSD adds lower levels of specification about how the processing is to be carried out. Design detail is added as elementary operations, which are given names and placed in a list. The list is numbered and elementary operations can then be assigned to components of the process structure as miniboxes on the diagram, as shown in Figure 3.6. Elementary operations should not be confused with elementary components of a structure and may be defined as follows:

- *Elementary operations*. Low-level design details of a process which are appended to components of the structure.

Elementary operations contain detail associated with a particular action and are a mixture of calculations and algorithms derived from the specification and computer-related operations included during implementation. In JSD detail may be added at any stage during specification;

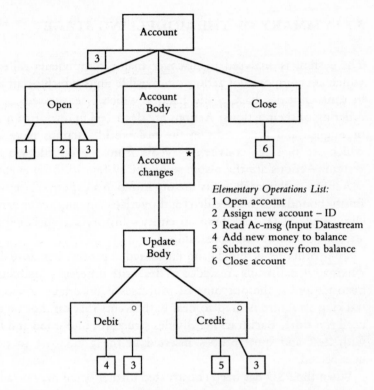

Figure 3.6 Process structure diagram showing elementary operations.

most of the detail is generally added later during the implementation stage, however.

Examples of elementary operations are Read/Write operations, computer-related tasks such as opening and closing files, and detailed specification of algorithms to be associated with actions. For instance, the mathematical formula for calculating compound interest would be an elementary operation to be associated with a Calculate interest action. When defining elementary operations in the modelling stage, however, analysis in terms of physical computer operations should be avoided because the specification is a logical statement of what should happen and not a statement of how it should be carried out.

In JSD, structure text may be elaborated in parallel with PSDs as the specification progresses. The initial structure text will show only the process actions. The text is then elaborated by adding elementary operations for reading and writing messages, a necessary part of the next stage of JSD, network analysis.

3.7 SUMMARY OF THE MODELLING STAGE

The system is analysed to discover major components called entities which are composed of actions, indivisible units which occur at a point in time. Entities have a life history which is expressed as the time ordering of their actions. Actions are described in one or two sentences of English and must exist in the real world. Actions have attributes which are message-conveying events from the outside world to the system. Actions are the response by the system to those events.

Actions within each entity are organized in an approximate sequence. Entity boundaries are decided, and overlaps and any redundant actions eliminated. Data describing an entity's history are specified as entity attributes with update constraints.

Time ordering is refined and illustrated in process structure diagrams. Alternative pathways are added to deal with different possibilities in life histories and if the outcome of processes is uncertain, a special back-tracking structure is used to deal with events that can not be predicted until run time. Backtracking divides structures into good and bad parts with 'test and quit' probes inserted in the good part to detect the unexpected.

When the PSD has been constructed further detail may be added to it as elementary operations which specify processing particulars.

Key points

- Entities model objects in the real world.
- An entity is composed of actions which model events in the entity's life history.
- An entity may have two or more roles, that is, different life history views of the same object.
- Action attributes are the messages from outside the system which trigger the action.
- Entity attributes are internal data of the entity which are updated by the entity's actions.
- Time ordering describes all possible life histories, not the most probable or desirable.
- Life histories that have an uncertain end are modelled by backtracking.

Key steps: Entity modelling

(a) From analysis notes, interviews and documentation draw up a list of actions which may be found in phrases, verbs and nouns.
(b) Describe actions with one/two sentences of English and list action attributes.
(c) Allocate actions to entities.
(d) Create preliminary action sequence in entity life history.
(e) Review entities to check
    (i) entity boundaries are correct,
    (ii) no separate lives exist (marsupial entities).
(f) Describe entity life history as a model process with a process structure diagram.
(g) Define entity attributes, the type and updating actions.

4
JSD NETWORK STAGE
Initial model phase

This chapter covers the formation of the initial model, that is the description of inputs and outputs of model processes. The system specification diagram (SSD) is introduced: this models the system as a network of the interconnected processes. The principles which JSD uses to connect processes together are described before the system specification is elaborated in Chapter 5.

4.1 SYSTEM SPECIFICATION DIAGRAMS

The JSD system is a network of communicating concurrent processes, that is, all the processes are running notionally at the same time. Processes are linked together by communication channels along which messages flow. The system specification diagram (SSD) records the network of processes within a system and how they communicate with each other. The SSD is a top-level view of the system which is not decomposed into subnetworks, etc: further specification is to be found in the PSD of each process shown in the SSD. There are four SSD components: model processes, function processes, datastreams and state vectors.

Figure 4.1 Datastream connection between processes: orders (ord) are read by the Order process which writes issues (iss) to Stock.

PROCESSES

Processes are shown as a rectangular box, and may be model or function processes (described in Chapter 5).*

DATASTREAMS

Datastreams connect processes together. Input datastreams should be connected to the left-hand side of the box and output datastreams to the right-hand side. A datastream is a message channel with a buffer organized on a first-in/first-out (FIFO) basis. Datastreams enforce a controlled communication link between processes, because if no message has been produced by an upstream process, by definition the downstream process is blocked while it waits for the arrival of the message. The datastream ID is placed inside the circle which shows the link (see Figure 4.1).

STATE VECTORS

State vectors (SVs) are the alternative process connection to a datastream, and are shown as a diamond on the SSD (see Figure 4.2). SVs are not discrete messages but data which describe and which are owned by a particular process. The state vector link is like an inspect or look-up instruction and does not enforce a controlled communication link between processes. State vectors take their identity from the process that owns them because they record the history of a process in data. State vectors are derived from the entity attributes. Information about system processes is accessed by state vector links.

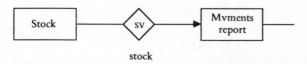

stock

Figure 4.2 State vector connection between processes: Mvments-report inspects the Stock process's state vector and writes a report showing the stock movements.

Arrows on the connectors show the direction of datastream movement, i.e. data come into the system generally on the left-hand side of SSDs

---

* In previous versions of JSD processes had the suffix 0 added to represent processes outside the system and suffixes 1 or 2 to represent processes within the system.

and progress through the system to output on the right-hand side. There is no strict convention, however, and datastreams can progress from the top to bottom of diagrams. It is important to note that the SSD is only a map; it shows the sources and destinations of datastreams and which processes are inspecting other processes' state vectors, but no information can be shown about when or how quickly processing occurs. Connections between processes may diverge and merge but datastreams must not divide between processes: once written, data-streams retain their identity until read. Connections can be marked to show the number of instances of each process type at either end of the connection, as illustrated in Figure 4.3. This is shown by a double bar on the connector at the many-occurrence side, whereas single ocurrences have no bar.

The connections between processes are described in more depth in the following sections.

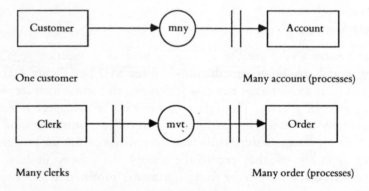

Figure 4.3 Notation on SSDs to show multiple instances of a single process type.

4.2 PROCESS NETWORKS

JSD views a system as a set of concurrently running processes which communicate with each other by messages. The two types of interprocess communication provided, datastreams and state vectors, differ in the relationship they create between communicating processes. Datastreams enforce a close interdependence because they are defined as a message stream with an exact Read–Write relationship. This means that every single datastream message which is written by an upstream process must eventually be read by a downstream one. State vectors, the second type of connection, do not enforce close coupling between processes.

The two types of JSD connection fulfil different functions within systems. Datastreams are used primarily for passing event messages (i.e. data messages that processes have to respond to) through the system, whereas state vectors are employed to access information about the system to create reports and query screens. The choice of which type of connection to use is left to the designer's discretion, but JSD does give guidelines as to the properties of connections and the context in which each type should be employed.

4.3 THE INITIAL MODEL

The initial model makes little use of interprocess connection. Its purpose is to show the model processes identified during the modelling stage and to record their inputs and outputs (see Figure 4.4). Model process connections at this stage will invariably be datastreams, that is, messages from or to the outside world which carry information about events to which the system processes have to respond.

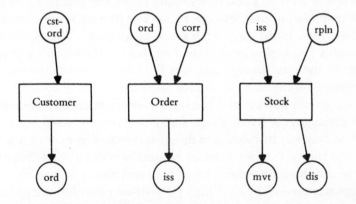

Figure 4.4 System specification diagram: the initial model, showing inputs and outputs to model processes with minimal connectivity between processes.

Input datastreams can be identified from action attributes. Each action in the model process should be inspected to determine if it consumes input or produces output. Action attributes of input actions will be conveyed in a datastream. The main question is how many input datastreams are necessary. To determine this the sources of events outside the system should be investigated. Separate sources will generate separate datastreams; otherwise event messages of different types may be grouped together in one datastream – for example, debit and credit messages from a customer going to an Account process. Sometimes event messages may have to be separated into different datastreams if there is some interdependency in their processing.

Little connectivity between model processes is specified at this stage; this will have to wait until the system is elaborated in the next part of the network stage (in Chapter 5). Before elaborating the network, however, it is important to examine the concepts JSD uses for interprocess communication in more depth.

4.4 DATASTREAM CONNECTIONS

Datastreams are potentially buffered message channels which act as a first-in/first-out (FIFO) queue. Each message written by an upstream process must be eventually read by a downstream process in the order in which it was written. For an illustration of the idea consider two clerks with a single in/out tray between them. The first clerk writes messages and places them in the tray which acts as his out-tray. The second clerk, who works a little slower than the first, has to take messages from the bottom of the tray because that was the order in which the messages were written. If the first clerk never takes a break the second clerk will never empty his in-tray, and messages may accumulate *ad infinitum*.

The timing of the Read–Write relationship is not defined because the two processes may be running at different speeds; hence communication is asynchronous. If a downstream process is waiting for a datastream message and there are none in its input queue, then it will be blocked at the Read statement. The blocking of datastreams may be used to control process activity and as a synchronizing mechanism between processes by switching processes on and off according to the supply of messages.

There are few constraints in JSD as to what should constitute one datastream as opposed to two or more. The definition of a datastream is usually determined by the events which it has to communicate. If events come from separate sources then a separate datastream is required for each set of event messages. Messages within datastreams may be heterogeneous (e.g. credit and debit records) or homogeneous, and may be fixed or variable in the quantity of data contained in a message. Datastreams may be used for control messages, i.e. as time markers or control flags. Datastreams used for this purpose should be separate from other streams because the JSD constructs of reading control (merges) will be necessary to specify how the process control should work.

The decision of what to place in a given datastream will be guided by the types of event which are contained within the datastream and how the messages are to be processed. In some cases the reading sequence may be important, particularly in heterogeneous messages with two separate components which have to be treated separately. The two parts of the message may have to be processed at separate speeds and therefore will need to be read separately from different datastreams rather than from one. Once written, datastreams may not undergo spontaneous subdivision between processes; a process may output several separate datastreams, however.

The Read and Write instructions for datastreams are appended to the structure diagram as elementary operations and may also be placed as Read(ds) and Write(ds) statements in the structure text. Care has to be exercised to ensure Reads and Writes are placed in the correct part of the process, i.e. Reads need to be placed before the actions which require data and Writes after actions which produce the data. The usual pattern is to place a Read immediately after an action to replenish the input with the next message, e.g.:

```
A-Process
    Read 1st;
        B-Seq
            Do X;
            Read next;
        B-End
        C-Seq
            Do Y;
            Read next;
        C-End
A-End
```

Datastreams are used not only for connections between processes but also for connections between the outside world and the system; hence input data for validation and system output, reports and displays are generally modelled as datastreams.

Timing and synchronization messages also use datastreams to ensure that the receiving process gets the message; in fact, in any situation where the designer wants to ensure that the receiving process gets a message, a datastream should be used. In this manner datastreams can be looked on as the 'registered post' of the system. Use of datastreams enforces close coupling between processes which in some circumstances may be undesirable. For these circumstances JSD uses state vector connections. However, the main reason for choosing between the two types of connection is not control but the view which needs to be taken of the data:

■ If a long-term view of events over time is needed, then use a datastream.
■ If a short-term snapshot of data is required, then use a state vector.

Datastreams are pipelines that convey messages describing events; these events may be parts of a life history or simple control messages.

4.5 STATE VECTOR CONNECTIONS

State vectors do not enforce any coupling between processes and are not written in an explicit sense. Conceptually state vectors are best thought of as a process instance's own data, that is, data items which pertain to its state, will identify it uniquely and will describe its properties. The concept is integrally linked with multiple occurrences of one process type, as each occurrence of a process has its own state vector to describe it. The state vector exists as long as the occurrence of the process does, and changes as its actions are executed.

The state vector is derived from entity attributes and can be thought of as data containing the history of a process, with data items recording the results of its actions. One of the recording items, called the *text pointer*, records the position in the process text (and by implication in the PSD) that the process has reached at a point in time. The role of text pointers assumes more importance in the implementation step.

State vectors are read by a Get(SV) instruction. Reads(Get SV) are entirely the responsibility of the reading process. The process whose state vector has been accessed is unaware of the action in the sense that the GetSV(A) instruction is a 'Spy on Process-A' type of operation.

Timing of a Get(SV) inspect can be important because it may be critical to ensure that the snapshot of a process's state is taken at a particular point in time. The process being inspected may be running slower or faster than anticipated, however, which means that the state vector may contain unexpected information. Worse still, the information may be assumed to be correct because of the assumption that the inspected process 'shouldn't have changed state by time t'. Timing of Get(SV) will also be governed by the speed of execution of the inspecting process, and if the inspected process has to be accessed at a specific point in its processing cycle (e.g. before and after an update) then timing can become critical.

State vectors are a useful type of connection for accessing information about the system, and allow great flexibility in implementation. Although the uncertain nature of timing may seem to be a problem at the specification stage, timing constraints can be added at implementation.

The use of state vector connections is common when data is required for reports and displays which are produced by function processes, described further in Chapter 5. The information is actually produced by the processes within the system and made available as they update their state vectors. To give a concrete example: if a report requires data about every account balance in a banking system, then it needs to inspect each occurrence of the Account process state vector. The Account process alone is responsible for updating the balance item in its state vector as it processes debits and credits.

State vector connections are used when no enforced Read–Write relationship is necessary, or even desirable, such as when the inspecting process is an online query. The information may be needed at any time and state vector connections give the freedom to look at the data on demand. In this way Account Report would inspect the state vector of the Account model process to ascertain if it is in credit or not. Another use for state vectors is for static data which are used for reference purposes, such as look-up tables, parameters and reference files. In this case the host process may be outside the system boundary but its state vector can be inspected. Examples of this usage are input validation processes inspecting customer state vectors for customer numbers or device control processes looking up device parameters.

4.6 MERGING DATASTREAMS

Few systems consist of linear sequences of single processes interconnected by single datastreams. Inevitably messaging pathways divide
and merge to form network interconnections in a system. Network
connections increase the complexity of a system, because when there
are two or more datastream inputs to one process there are several
alternatives for the reading order: together, one at a time, or alternating
between the two. JSD views systems as communicating asynchronously
and hence the first (together) option is not used; however, the reading
order can be specified in several ways using design constructs called
datastream merges. The merge construct does not exist with state vectors,
since the reading order is dictated by the internal structure of the
inspecting process.

Datastreams are said to be merged when two or more streams are read
by one process. Whereas the writing order of multiple output datastreams has few constraints, reading of multiple inputs needs to be
controlled because of the possibility that one of the input channels may
be empty. By definition in JSD, if this occurs the process must halt and
wait for the input. Messages may be absent for several reasons; one
writing process may be running faster than the other, or one communications channel may transmit messages slower than the other. The
consequence is that the sequence of message arrival may not be
predictable.

If the sequence of message arrival is known, JSD uses a *fixed merge* to
read messages in a predetermined order. In simple cases two channels
may be read alternately, i.e. a reading sequence of ABABAB ... etc.
More complicated patterns can be specified, such as 'read five As then
two Bs', and so on. Message arrival is often not so well determined,
however. To get around the problem of uncertain arrival of messages
JSD uses a special construct called a *rough merge* to pool the contents of
datastreams. The choice of merge type is left to the designer. The
essential difference revolves around how predictable interprocess
communication is going to be, which can be summarized in the
question: 'Can the arrival sequence of messages on the two (or more)
datastreams be exactly predicted or not?'

If it can, then a fixed merge (see Figure 4.5) is used. This specifies the
exact order of reading, with the consequence that if one of the
datastreams is empty the process will wait for eternity if necessary until

the next message arrives, even if data is accumulating in the other datastream. A sample structure text for a fixed merge is:

```
Process Fixed Merge Seq
    Main-Proc Iter
        Read(Stream A);
        Do Action A;
        Read(Stream B);
        Do Action B;
    Main-Proc End
Process Fixed Merge End
```

Absence of an A message will cause the process to wait at the Read(Stream A) instruction. The processing of fixed and rough merges becomes embedded in complex PSD structures, so fixed merges can be used to halt process execution at several points in a process life history.*

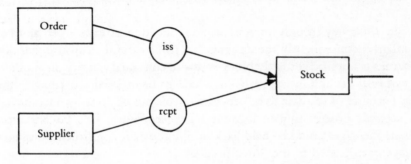

Figure 4.5 Fixed merge in a process input: in this case issues and receipts arrive at Stock in a preset order, i.e. one issue is followed by one receipt.

Rough merges are a form of 'inspect datastreams and read' instruction in which the reading process examines the pooled contents of two or more streams and processes the first message it encounters. The downstream process sees a rough merge as a message pool which still obeys the FIFO queue rule, so in the example shown in Figure 4.6 the Stock process sees a message pool composed of two different types of message in the order of writing, e.g. iss, iss, iss, adj, iss, iss, adj. If one

---

* Previous versions of JSD have variants of the fixed merge called *data merges* and *periodic collates* which allowed for more flexibility in the downstream processing of data. In this case messages must still be present on both datastreams, but the type of processing depends on qualities of the data. Absence of a message on either stream would cause the process to halt and wait, possibly indefinitely, for the arrival of the next message. Data merges, therefore, had little advantage over fixed merges in dealing with uncertain arrival of data.

stream has many messages it will receive preferential treatment while
the stream with fewer messages will receive proportionately less attention. If this becomes a critical disadvantage in the design then a process
has to be introduced to control the merging order.

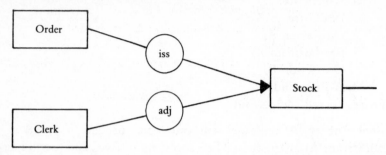

Figure 4.6 Rough merge in a process input: the order of message arrival is not
predetermined. Issues and adjustments go into a single pool and Stock has no way of
knowing whether it will get an issue or an adjustment next.

In this way rough merges can favour one stream over another
unfairly; consequently the designer has to be careful to ensure that one
stream is not inadvertently blocked due to continual activity on another.
As a result rough merge reads may have to be ordered according to the
importance of the data to be processed. This can be done by introducing
a separate process to prioritize messages (see Figure 4.7). For instance,
issue messages could be held back until important adjustment messages
have been passed to the Stock process.

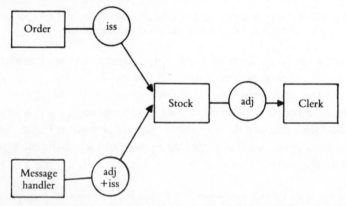

Figure 4.7 Prioritization of a rough merge by introducing a message handling
process: this accepts adj and iss messages in the order of appearance and then feeds
adj and iss messages to Stock in a predetermined order. Buffering of messages is
provided during the implementation stage.

Ordering may have to take into account the rate of data production upstream. Sometimes data may be produced infrequently but nevertheless it is important – in this case it is necessary – to interrupt a busy datastream to allow priority processing to occur. Prioritization can be effected by placing marker records in the busy datastream at specific intervals to halt processing and allowing the infrequent stream to be read. Another method is to create a separate fixed merge of the busy stream with a time marker datastream. The time marker controls when the busy stream is switched off and on.

Rough merges are shown in structure text as follows:

```
Read(Stream A and Stream B);
Action Sel(A message present)
    Do Action A;
Action Alt(B message present)
    Do Action B;
```

Read and Write instructions are added to PSDs as elementary operations; they have to be placed with some care, however. Generally Read instructions are placed before the action which consumes the data, and when the action is in a loop it is good practice to place another Read immediately after the action to make sure the input data is replenished for the next cycle. Write instructions should be placed after the actions which produce the data but the location is less critical; for instance, a heterogeneous datastream may have a Write placed after several actions which all contribute data to it.

4.7 SUMMARY OF THE INITIAL MODEL PHASE

System specification diagrams are used to show the system as a network of concurrently running processes with communication links. SSDs have three main components: processes, datastreams and state vectors. Processes are connected by either datastreams or state vectors. The initial model shows a minimal view of the system network with model processes and their datastream inputs and outputs. Its purpose is to record the system input and output message streams which convey events to and from the system; apart from that little connectivity is shown. Datastreams have the following properties:

- They are unbounded buffers acting as FIFO queues.
- They enforce Read and Write relationship; messages must be read in the order in which they were written.
- They consist of data messages which must be consumed by the downstream process.
- They block activity of downstream processes if no messages are present.

State vectors have the following properties:

- They belong to a process occurrence, are updated by it, and represent its internal status as variables.
- They may be single items or records of data pertaining to a particular instance of a process type.
- They exist as long as that process exists and may be read or inspected at will.
- They are accessed on demand by the reading process.

State vector connections are used to obtain information about processes in the system for reports and displays or whenever there is an on-demand need for information. Datastreams are used to connect processes when:

- event messages are being sent or received;
- control messages which must be read are required;
- data is to be output as results, processed transactions or reports.

Care has to be exercised when merging multiple datastream inputs into one process. If the order of arrival of messages is known then a fixed merge is used; otherwise a rough merge is used to inspect each stream and process the first one encountered that contains a message.

Key points

- The initial model records system input and output messages to model processes.
- To choose between datastream and state vector connections:

  if a long-term view of many events is required use a datastream;
  if a short-term snapshot of a (process) state is required use a state vector.

- Datastreams record events; the number of datastreams depends on the sources of events and how events are to be processed.
- Choosing between fixed and rough merges depends on how certain the arrival of messages is. If no arrival order can be specified then a rough merge is used.

Key steps: Initial model

(a) Using action attributes, list the model process inputs.
(b) Examine model process actions for outputs.
(c) Determine which type of connection to use, depending on whether a short-term (snapshot) or long-term (event history) view is required. Use state vector connections for the former and datastreams for the latter.
(d) Construct the initial model, showing model processes with their inputs and outputs.

5
JSD NETWORK STAGE
Elaboration phase

This chapter deals with the elaboration of the system specification in JSD, in which the input and output subsystems are added to the initial model. The different types of JSD function processes are described, then methods of specifying timing are examined and JSD data analysis is reviewed. This completes the JSD network stage which delivers the logical system specification.

5.1 ELABORATION OF THE SPECIFICATION

At the end of the initial model phase a JSD specification is a minimal view of the system consisting solely of model processes and their input and output datastreams. Model processes are necessary to respond to the basic problem of the system, but little other detail has been included. In the elaboration phase other parts of the system are added as function processes which implement the information part of the system and the user interface.

JSD deals with the complexity of developing systems by specifying different parts of the system in turn. First the core of the system is modelled, i.e. the model processes which respond to events in the world and fulfil the system's main purpose. Only after this basic modelling has been carried out are the other main system components added – the input and output subsystems. The output subsystem is responsible for providing information. This is the part of the system which provides data about the performance and status of model processes, known in more familiar terms as report programs and query screens. Finally the input subsystem is specified to complete the model.

Jackson draws a clear distinction between model processes and other processes, called *functions*. This distinction emphasizes the separation of the modelling of the system's more important components from the expansion of the model with ancillary processes. As specification progresses different parts of the system are added, effectively partitioning it into subareas of model/event processing, information processing and user interface (including validation). By enforcing this view on systems JSD anticipates the effects of maintenance.

Central model processing is likely to be the most stable part of a system; it relates to the basic entities which, in turn, reflect the basic business of the organization. Jackson argues that if these entities have been correctly modelled then they are likely to change infrequently, if at all. Modification of model processes, although not impossible, should only be caused by major changes to the basic operation of the system and generally reflects a change in the business environment.

Functions, in contrast, contain the changeable part of the system. Information reports, query screens, validation criteria and the user interface are all liable to more frequent change which, in Jackson's view, is driven from people's information requirements about processes rather than the need to change basic processing itself. By partitioning the more volatile parts of the system into separate processes, JSD makes modifications easier to carry out, and limits the unpredictable effects of change. When changes to a report or an input validation procedure are requested the analyst knows where the change should be located within the system.

5.2 ADDING NEW FUNCTIONS: EMBEDDED OR IMPOSED

There are discrete rules which define how functions may be added to JSD systems. The basic rule states that:

■ Only elementary operations may be added to existing model processes.

If anything larger has to be added (i.e. a new action or group of actions), then a new function process must be created and linked to the rest of the system. The reason for this is to preserve time ordering in model processes: any disturbance to this would corrupt the results of earlier analysis.

Trivial changes such as addition of Read and Write instructions can be added to model processes by appending an elementary operation. No change occurs to the PSD components because, by definition, no new components can be added. In JSD such trivial changes are called *embedded functions* in the sense that they are embedded within an existing model process. Larger changes are called *imposed functions* because change is imposed or added to the system as a new process. For example, in a banking system an Account model process may be modified to produce a list of all credits to accounts by a simple addition of a Write elementary operation to the Credit action. If a more complex information requirement has to be satisfied then an imposed process will be needed. Thus if a report listing all credits above £1000 were required then a new action would be necessary to select credits above £1000; consequently this action would have to be placed in a separate imposed function. The process connection between model process and imposed

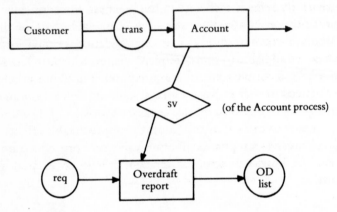

State vector connection

AccSV	Owner:	Account
	Access Path:	ascending order by acct–ID
	Selection:	balance < 0
Timing	Report production on line; display within 5 seconds of request	
SV updating	Balance updated accurate at start of working day, maximum lag 24 hours	

Figure 5.1 Addition of an imposed function to an SSD: the function process, Overdraft-report, inspects the Account process state vector, determines if account(*n*) is overdrawn and outputs a report on datastream ODlist.

functions is by a state vector, e.g. the connection between Account and Overdraft-report would be by inspection of the credit balance in the Account state vector (see Figure 5.1).

State vectors are used because the function process invariably needs to take a snapshot of the model process's state, e.g. accounts status at the end of the month or week or on demand. The state vector connections between imposed functions and model processes are reports on data owned and updated by model processes. Those data will be held in the model processes' state vectors, which can be inspected at will. Loosely coupled connections by state vectors are useful as many reports and query screens require information on demand. Use of datastream connections would be inappropriate because these convey a longer-term view as a stream of events. In addition the Read/Write constraints of datastreams could impose an undesirable timing constraint on the function process while it waited for the model process to transmit the necessary data. Care has to be exercised with state vector connections to function processes, however. The main problem is ensuring that information is captured when the designer expects it to be, and therefore due regard has to be taken of a model process's probable current position in its time-ordered sequence at the time when its state vector is inspected.

An illustration of the problem is the Account process, which in this case has three variables. One accumulates the number of credits given to a customer's account, one accumulates the total amount of credit, while the third holds the result of calculating the average value of credits. The report function wants the total credit value, number of credits and the average value of the customer's credits. The Account process has two actions:

Add action
  Add NewCredit to TotCredit giving TotCredit
  Increment Number of credits by 1
Calc-Average action
  Divide TotCredit by N credits

If the state vector inspect occurred between these two actions then the average would be in error. There is no infallible way of ensuring where a process is in its life history, but use can be made of a process's text pointer to determine when actions have been completed. The state vector inspect can then be sure that the action it is interested in has been completed, although without a lock mechanism it cannot tell whether the next cycle has already begun.

When information functions are specified the access path of the state vector is stated. This may be a simple access to the state vector of a model process, or a more complicated path which requires inspection of different state vectors. The timing requirements for the production of output and how up to date the information produced must be are also noted. The accuracy of the information will depend on how frequently the state vectors are updated by their host processes.

5.3 INTERACTIVE FUNCTIONS

Interactive functions are encountered briefly during early analysis (see Chapter 2), as procedures and tasks with short life histories. A more succinct definition of interactive functions is:

■ *Interactive function*. A process which creates input to a model process when that input is not immediately derived from outside the system.

In more concrete terms, interactive functions are separate processes implementing procedural types of processing. These functions arise in three ways: as parts of the system discovered during entity analysis, as parts of the user interface (see Section 5.4) and as extensions to the system's processing added to satisfy user requirements. Computer-related activities such as backup, recovery, restore and file maintenance will also result in the addition of interactive functions to the SSD.

Interactive functions may have been discovered during initial analysis, as algorithms and pieces of logic interactive with model processes to complete a specific task. To illustrate an interactive function found during analysis, consider an insurance system. The initial model has a process Claim which models the life history of a policyholder's claim for an accident. There is also a complex algorithm which assesses liability and determines how much the company will pay out. The algorithm would be an interactive function Assess-liability.

Interactive functions are connected to model processes with data-streams, because they form an integral part of the model system processing. Interactive functions contribute event streams to model processes and a long-term view of their input is needed. Model processes have to respond to events output by interactive functions within the system in a mandatory manner; consequently datastream connections are necessary to ensure that messages are read and dealt with.

5.4 FILTER PROCESSES AND THE USER INTERFACE

Some aspects of data validation may have already been described in model processes, but input entry poses separate problems which are no concern of the model process. JSD reserves a specific part of the system for dealing with the issues of data validation in the network stage of development.

Allowance has to be made for input data being good, bad, corrupt or correct but mistimed. In the elaboration phase problems of data validation are addressed directly by introducing new interactive functions called *filter* processes to protect other system processes from the consequences of bad data. In addition to their validation duties, filter processes also provide the user interface.

First it is necessary to consider how data may be invalid. There are three categories of bad data which must be guarded against and a fourth against which it is difficult to take countermeasures.

(a) *Corrupt.* The input data are corrupt in the sense that they are not decipherable. This may occur with errors in electrical signals during data transmission over a telephone line. The corrupt data have to be discarded and a request to retransmit given.

(b) *Invalid.* Data are decipherable but of the incorrect type or range. This is the normal sense of validation.

(c) *Incorrect sequence.* Correct data arrive but at the wrong time or they are not the expected event. Either they must be buffered until required or an error message issued.

(d) *Mistaken (valid).* The data are valid and arrive at the right time but they are not what the user intended. This condition cannot be detected without a considerable knowledge of the user's intentions, however. The user interface should allow the user to backtrack and retrieve the situation.

Filter processes have to protect system processes from the effects of invalid, corrupt and incorrect sequences of data items. Some validation may have already been described in other system processes. This validation should be removed and placed in filter processes.

Input validation deals with uncertainty at run time, i.e. we have no way of knowing whether data will be correct or not until they arrive. Filter processes, therefore, are often backtracking structures because this is the JSD construct specifically designed to deal with run time uncertainty. The first step is to model the arrival order of input datastreams, and this will become the Posit side of the process structure. Next validation tests are added for each input and Quits inserted into the PSD to deal with validation failures. Each Quit must lead to the Admit branch where in most cases further action will have to be taken, such as error messages or undo options, although occasionally no further action may be necessary (the side effects are beneficial or neutral). The PSD Admit branch is elaborated to include actions which deal with undesirable side effects.

Filter process should validate the sequence of input messages, to deal with type (c) errors. In this role they are referred to as *context filters* in JSD. In an order-processing system, for example, the context filter would be responsible for ensuring that orders were not amended before they had been received, or dispatched before credit clearance. This would involve accessing model processes' state vectors to ensure that the input message is of the type which is expected. Alternatively the

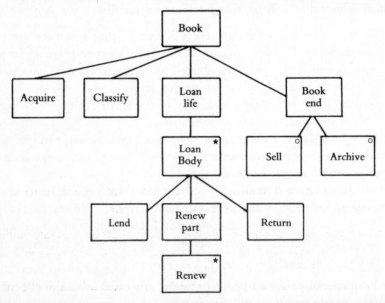

Figure 5.2 Book model process. [From course notes (unpublished) of Michael Jackson Systems Ltd, *JSD for Information Systems: Basic Course*, MJS Ltd, London.]

filter process can maintain a state vector recording the input event history and communicate with the system process by datastream messages. A model process and its corresponding context filter are illustrated in Figures 5.2 and 5.3.

Type (d) errors cannot be detected because they appear to be perfectly reasonable events although unfortunately not what was intended. To deal with these errors corrective actions can be added to filter and possibly system processes. Corrective actions allow the user to undo an action and correct mistaken input; examples are the undo option in a word processor to go back to the previous state before the last operation, or a de-allocate action in an order processing entity with a Place–Change–Allocate–Deliver life history.

So far a filter process has been specified for only half of the task (i.e. event validation); to complete it the user interface and dialogue control have to be added. A separate process is added to handle the user interface and simple type (a) and type (b) errors. These simple filters are based on the action attributes, which have to be validated. Actions should be grouped together into transactions according to the user's view. The transactions will then become the menu and form-filling screens of the system. To illustrate the grouping process: consider a library system having the actions Lend, Acquire, Classify, Remove, Renew and Return. In the user's opinion Acquire, Classify and Remove, are all about book stock management and hence should be grouped into one menu, while Lend, Renew and Return are all loan-related and should be grouped into a separate transaction.

Once the simple filter process has been defined on this basis it can be elaborated to add extra actions for:

finding attributes not supplied on input, e.g. Customer ID and address, and other reference information;
policing undesirable events, e.g. >10 changes to an order;
user guidance and control, e.g. escape from option entered by mistake, undo last event, access to help screens.

Interface and user–computer dialogue design are a separate subject in their own right and the reader is referred to Monk (1985) and Shneiderman (1987) for further details. JSD is quite useful for modelling the human–computer dialogue, however, so some advice about using JSD for interface design will be given.

The starting point is to model the time ordering of the user's actions when operating the computer system. The dialogue can be modelled as

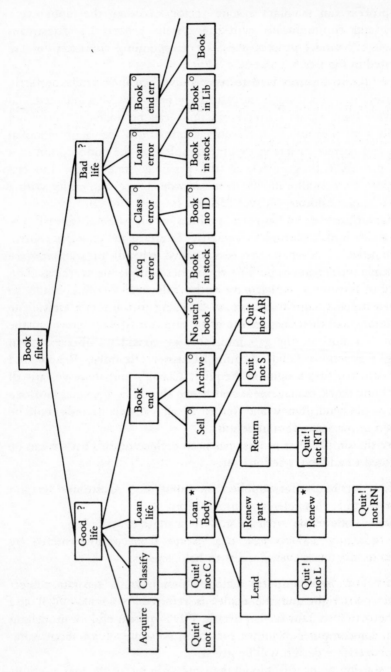

Figure 5.3 Book context filter: this checks for type (c) errors (events in the wrong order).

a series of steps in which the computer asks a question and the user replies. Some of the data entry dialogue will already by modelled in context filter processes, but most of the system control dialogue will be added in simple filter processes. When analysing the user–system dialogue a useful heuristic is to ask the following questions at each step:

- Could the user want to escape (i.e. quit the operation) at this stage?
- Could users require to backtrack at this point to correct an error? If so, where should they restart?
- Does the user require help at this stage?

Answers to these questions will cause actions to be placed in the filter PSD as Quits for escape, more selections for help and Quits linked to Admit branch actions for undo or backtracking functions. In this way the structure is elaborated by tracing through all the input stages and modelling the user's actions at each step. The filter process becomes a composite model of user and computer actions in a time order. An example of filter process elaboration is shown in Figure 5.4. First the input validation is modelled, and then the user control actions are added to allow the user to either enter, edit or escape from the data entry sequence.

Another step is necessary to ensure that the user interface is properly implemented in filters. Actions have been added to give users control over the interface, but so far no prompts or feedback messages have been incorporated in the design. When elementary operations are added to the PSD, messages should be added to ensure users get correct and meaningful feedback messages from the computer. A guideline to follow is to check the following conditions at each input step:

- Is an input prompt necessary before the data entry step to tell the user what to do?
- Has an error message been added for the invalid data?
- If there is going to be a delay in response time, has a message been added to inform the user?

Addition of messages in the appropriate places in the filter processes will provide feedback for users to keep them informed and reassured about what the computer is doing. A simple filter process PSD is shown in Figure 5.4 and a sample SSD for part of an input subsystem is given in Figure 5.5.

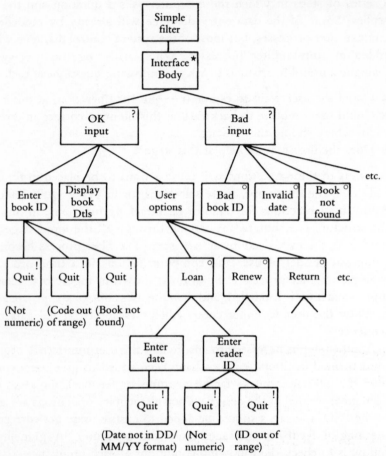

Figure 5.4 Simple filter for Book process: this checks for type (a) and type (b) errors in the input and implements the user interface. The reader-related tasks of Loans, Returns and Renewals have been grouped together; another filter would validate book management tasks such as Acquire, Classify, etc.

5.5 TIMING AND SYNCHRONIZING PROCESSES

Timing merited a step of its own in Jackson's original definition of JSD, but is now included within the network stage. Two problems have to be addressed: timing of actions within processes and synchronization between processes. To achieve timing JSD uses special messages called *time grain markers* (TGMs), which act like datastreams but contain timing

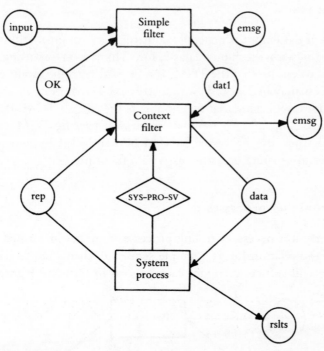

Figure 5.5 Filter processes showing SSD connections. The simple filter checks the input for type (a) and type (b) errors and then passes it on to the context filter which checks for event sequence errors. If no errors are found the input is passed to the system process. Datastream connections are used to communicate feedback about the results of validation on the rep and OK datastreams.

information. In complex synchronization problems new timing processes are introduced into the system. The SSD is updated to show addition of timing.*

Processes within JSD systems are activated by the presence of data, and thus process timing and activation can be controlled by the presence or absence of messages on datastream connections. Processes may also be connected by state vectors which enforce no timing dependency. This unpredictability may be acceptable in some cases, but often there is a requirement to run a weekly or daily report or to read a datastream at the start of each day. As JSD processes are concurrent and long-running, such timing constraints require TGMs to control stopping and starting processes.

* In previous versions of JSD, the updated SSD was referred to as the system timing diagram (STD) and timing was covered in the 'System Timing Step'

Timing reads

Timing of data input is generally controlled by rough-merging a TGM datastream with the input datastream. The reason for using a rough merge is to ensure that the TGMs can be read with the input; otherwise the process may be blocked awaiting its next TGM. Note that the rough merge does not ensure that the input gets the correct TGM. If a one-to-one relationship of each input message to a specific TGM message is required, then the processes have to be controlled by synchronizing messages in separate datastreams with a fixed merge.

Synchronizing processes

Frequently it is necessary to link processes together in a timed relationship. If the relationship is a simple sequence, then a single datastream message will suffice to trigger each process in turn (see Figure 5.6).

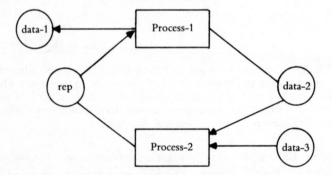

Figure 5.6 Use of datastreams to synchronize processes. Process-1 controls the activity of Process-2 by sending it one data-2 message at a time. It can check on what happens to each message because Process-2 has to write a rep message after it has consumed each data-2 message. The rep datastream controls Process-1 by a fixed merge, forcing Process-1 to wait for a rep message. The processes have a conversational datastream connection which controls interaction as a two-way conversation, e.g. Process-1, then Process-2, Process-1 and so on.

The first process has a fixed merge of its input datastream with a trigger datastream. Activity is triggered by the first rep message. Process-1 will then wait until it receives the next rep message before it can consume more input. Use of datastreams enforces a Read–Write relationship between processes; the designer can therefore be sure a message must arrive before an action is triggered.

Clock programs

When many TGMs have to be produced, addition of a separate process may be necessary to control their production. Single TGMs may be input from outside the system as simple datastreams, but several TGMs or TGMs with complex interdependencies should originate from a Clock program. Clock programs are also necessary where there are linked time dependencies between several processes. Consider the account update system shown in Figure 5.7: it consists of two processes, one of which (Deposit) accepts and validates Credits and Debits throughout the day and at the end of the day passes the resulting sum of transactions for each customer through to the second process (Balance), which adds transactions to the balances and prints a report.

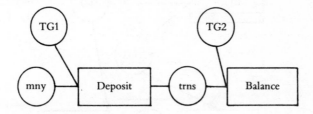

Figure 5.7 Use of time grain markers to activate processes.

A simple implementation may be to use two time markers, TG1 and TG2. The first triggers Deposit to send the summed transactions to Balance and the second triggers Balance to produce its report. We might expect that this specification is sufficient, but JSD processes are unpredictable in their running speed. It may be that Deposit runs so slowly that not all the transactions are ready for Balance by the end of the day. Undeterred by this revelation we could specify that TG2 will be later than TG1 (e.g. 6.00 p.m. and 5.30 p.m.) to give Deposit a bit of leeway for slow running. But this solution is guesswork. We have little information on Deposit's running speed, so a far better solution is to synchronize the processes.

To achieve better synchronization we must ensure that Balance does not receive its TG2 before Deposit has completed its processing. To do this we employ a Clock process which monitors the consumption of TGMs and is responsible for producing TG2 only when Deposit has read all its TG1s. The structure text for the Clock process is:

    Dep-Clock Iter while true
        Write TGM to TG1 stream;
        Inspect TG1 stream;
        Wait until all TGMs are consumed;
        Write TGM to TG2 stream;
        Wait until consumed;
        Read NXTDAY from input DS;
    Dep-Clock End

The resulting SSD is shown in Figure 5.8.

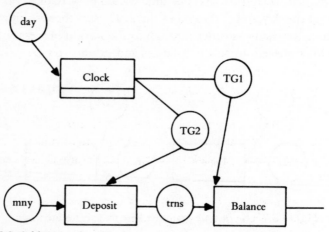

Figure 5.8 Addition of a Clock process to produce TGMs. Clock process is denoted by additional bar.

5.6 CONTROLLED DATASTREAMS

Many JSD processes have multiple occurrences of each process type; consequently individual instances are updated, created and deleted from the system. This creates problems if two or more inputs are destined for the same individual process; which input should be taken first – the create or the update? This is a replay of the familiar 'multiple update to one record' problem in databases. Some locking and unlocking mechanism has to be introduced.

    Another problem requiring locking is interdependence between two messages. Interdependency may have little to do with time; for instance, two items may be linked in the manner of 'Send B only if A is OK'. Some feedback message is required to say 'A is OK' before B is transmitted. One implementation is to use feedback datastreams but this is an inelegant solution and runs the risk of deadlocking the processes if the feedback message gets lost.

Both problems can be solved with *controlled datastreams*. These introduce the idea of controlled reading of two datastreams in a rough merge. The idea is to allow inspection of the fate of datastream messages, so an upstream process can inspect the fate of its own output. Another view of the concept is that controlled datastreams deliver a guarantee of the downstream process's status after reading a datastream message. Given this guarantee, message prioritization of two competing or interdependent datastreams can be specified.

Controlled datastreams have a state vector inspection by the upstream process of the downstream one. The effect is that while it is writing messages the upstream process can lock its datastream, forcing the reading process to read this stream alone; it then inspects the downstream process's state vector (note this inspect is not shown on the SSD in Figure 5.9) to determine when the datastream message has been processed, and finally it unlocks the datastream to allow a return to normal rough merge type reading. Three new elementary operations are provided:

- *Query(datastream)*. Ascertains status of datastream message by inspecting downstream process state vector.
- *Lock(datastream)*. Locks the reading channel to the nominated datastream. Messages on all other channels cannot be read until an unlock instruction is given.
- *Unlock(datastream)*. Releases the reading channel and allows messages on other datastream channels to be read.

To illustrate the use of controlled datastreams consider another bank account system, depicted in Figure 5.9. There are three processes: Account, Deposit Account and Transfer, a function which automatically transfers money from the deposit account to the current account if it goes into overdraft. Deposit Account also has normal withdrawals and deposits and must not be overdrawn. If a normal rough merge is used, however, the Transfer function may inspect Deposit Account's state vector, which shows a credit of – say – £500. Suppose it sends a transfer message for £300 to be sent to the current account. Before this can be effected an ordinary withdrawal of £300 is made from the deposit account, which would result in an overall overdraft. Transfer needs to stop Deposit Account from reading its normal deposit/withdrawal datastream.

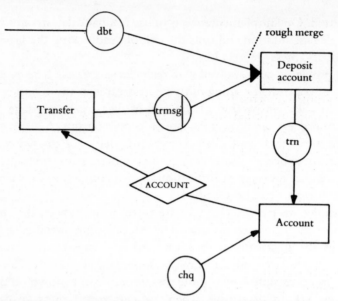

Figure 5.9 System specification diagram showing controlled datastreams: here Transfer can lock the datastream trmsg to make sure that Transfer messages are read before any more dbt messages arrive for Deposit account. The trmsg causes Deposit account to write a trn message to Account to top up the balance.

The structure text for Transfer would be as follows:

Begin interaction(Trans)
    Lock Trans;   initialization
    Query Trans;   (Get SV(Deposit Account, Trans)
                  – should be empty and waiting for a message)
    Write Trans;
    Query Trans;   (Get SV(Deposit Account) – finished with Trans)
    Unlock Trans; (Release Trans, allow Dep to be read)
End interaction(Trans)

The controlled datastream means that messages on the locked channel will be read and must be consumed and that the controlled stream must be unlocked before the second stream can be read. If the first message is incorrect for some reason then the Lock is not relinquished.

The essence of controlled datastreams is that processes lock message reading and determine the fate of messages by inspecting a downstream process's state vectors. In this way a double check can be carried out to ensure that a message once written (a) has been received and (b) has been processed.

5.7 JSD AND DATA ANALYSIS

At first sight JSD may seem to have little to say about data. Closer examination, however, reveals that JSD can help data analysis and has some very positive benefits for designers of data-oriented systems.

Data are first recorded in a JSD specification as entity attributes. These attributes will become fields in the state vector record which describes and is updated by the process which owns it. Because JSD links data with actions and views data as the result of change caused by actions, the method gives a good specification of update procedures and integrity constraints. JSD actions are, after all, doing the updating and PSDs will show the circumstances in which the updates can occur. These constraints can be built into update procedures in databases or file maintenance procedures.

To amplify the point, suppose that a library system used a database and it received various requests to Purchase, Catalogue, Borrow and Return a book: then a JSD specification can define the update constraints. From Figure 5.10 it should be apparent that a book at arrow 1 will have just been classified, and before this it could not be loaned; at arrow 2 a book out on loan cannot be loaned again before first being returned.

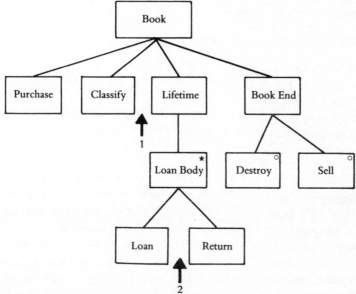

Figure 5.10 Book process PSD. [From course notes (unpublished) of Michael Jackson Systems Ltd, *JSD for Information Systems: Basic Course*, MJS Ltd, London.]

In addition to specifying update integrity constraints JSD, through the constructs of state vector inspection and function processes for queries and reports, gives a clear definition of data access requirements. These can be implemented as query language commands and procedures.

Logical data structure definitions can be created from entity attribute lists by the addition of a key field which will be the unique identifier of the occurrence of that process. Addition of action attributes from common actions includes foreign or link keys into the data structure according to the relationships established by actions. Hence in a library system the Loan action of Book may have attributes of Book ID, Loan date and Member ID, the last-named attribute forming a link key to the Member entity. Process state vectors will accumulate descriptor attributes and link keys to other entities from action attributes.

Process state vectors, so far, are not normalized, but during the process of JSD analysis elaboration of the PSD generally causes normalization of data at least as far as first normal form. Taking an order process: if we have a single product per order then the model process Order is correct. If we have many products per order, however, we will discover a marsupial entity which splits off the repeating group (Order-product) to maintain first normal form relations in the two state vectors.

Model process:Customer

Action: Amend(Customer entity)

Action attributes
  (Customer ID)
  (Order ID)
  Amended field ID
  New value

Entity Attributes (Order)
  (Order ID)
  Customer name
  Customer address
  Order date
  Delivery date

  Product code
  Quantity
  Unit cost

Total cost–net
VAT
Total cost
Order priority

The new entities and their respective entity attributes become:

Entity: Order

Attributes: Order ID
Customer name
Customer address
Order date

Entity: Order-product

Attributes: Order ID
Product code
Product quantity
Unit cost
Total cost–net
VAT
Total cost
Order priority

The effect of marsupial entities is to normalize data at least to first normal form. To create data structures from the collections of entity attributes it is necessary to add link keys to identify each new marsupial data structure back to the parent entity (Order). This would also be done in standard normalization. For the purists of normalization who wish to go further it is worth bearing in mind that the main purpose of normalization is to eliminate redundancy in a relational database so that data can be accessed flexibly in tables and update integrity is assured. JSD models what happens to that data when it is either accessed or updated. Thus if two items are not split up by JSD entity attribute grouping, it is probably because their values and access paths are fixed and so is their relationship, e.g. Town (Leeds) and County (Yorkshire) have a good chance of staying that way. JSD has modelled actions which define the relationship of the data items, making further normalization unnecessary.

JSD can also give guidance towards the relationships between entities. This is effected explicitly by links through common actions and by datastream links between actions in an implicit manner. If a Library

system has two entities Book and Library–Member which are linked by
the common action Lend, then there is an explicit data link between the
entities of the form illustrated in Figure 5.11(b).

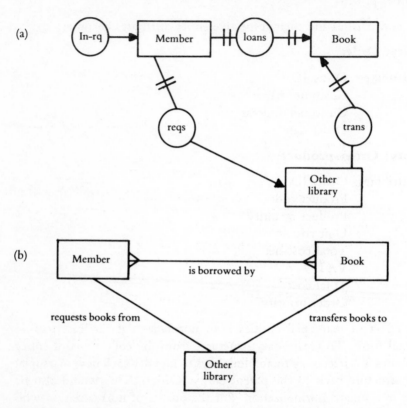

Figure 5.11 (a) SSD and (b) entity relationship diagram of part of a Library system.

The relationship is of the many-to-many type, a fact also recorded in
the SSD diagram shown in Figure 5.11(a). Other actions and datastream
links can then be examined to see if they also represent relationships that
can be illustrated in an ERA (entity/relationship/attribute) diagram. In
the case of the library an interlibrary loan request would cause a further
relationship to be formed between Other-Library and Book.

In summary: state vectors of JSD processes are produced from the
collection of entity attributes. The discovery of marsupials within entity
structures should have created normalization to first normal form in the
state vectors. To convert the state vector into a logical data structure,

extra attributes related to common actions between entities are added as link keys. Some further normalization may be necessary, but the update and access requirements should be studied using PSDs to decide if further normalization is justified.

5.8 SUMMARY OF THE ELABORATION PHASE

The initial model of the system is elaborated to include processes which provide information (reports and inquiries), input validation and the user interface. JSD has specific rules for adding modifications to the initial system model. Minor changes at the elementary operation level may be added to existing model processes (embedded functions); any more substantial changes are placed in new function processes (imposed or interactive functions) which are connected to the existing system.

Filter processes are a type of interactive function which implement the user interface. As input validation and the user interface have to deal with uncertainty at run time, filters are usually backtracking structures.

Timing constraints are added in this step. Timing is important to synchronize processes and to trigger actions within processes. JSD uses special datastreams, called time grain markers (TGMs), for this purpose. TGMs are rough-merged with ordinary datastreams to control the arrival of messages and the execution timing of processes. Synchronization can also be achieved by exchange of datastream messages between processes. For complex timing a separate Clock process is introduced to control production of timing signals. Interdependency of messages can be specified by controlled datastreams which allow reading locks to be placed on channels to ensure the reading order.

JSD data descriptions are produced by collating entity attributes into process state vectors. The analysis of entities for marsupials produces normalization of data structures to first normal form as a by-product. JSD specifications are useful for data analysis and controlling the integrity of database updating.

Key points

- No change is allowed to model processes apart from addition of elementary operations.
- System output is created by function processes which are connected to model processes by state vectors.
- System input is validated by input processes.
- Dependencies between datastreams and reading priorities can be specificed by controlled datastreams.

Key steps

(a) Elaborate the system network by adding functions. Most functions will be new processes, although trivial ones may be embedded within model processes.

(b) Add connections between function processes and other system processes. These connections will usually be state vectors.

(c) Add any input and output datastreams to functions to trigger their activity (input) and convey their output.

(d) Add interactive functions to the system network. Connect interactive functions with datastreams.

(e) Specify context filters to handle event sequence errors for system processes.

(f) Plan the user interface by specifying system input from action attributes. Group input into transactions according to the user's view.

(g) Design simple input filters to validate input and implement the user–system dialogue.

(h) Add timing constraints to the system network as time grain markers, using Clock processes for complicated cases.

(i) Use controlled datastreams where there is interdependency between the reading order of datastreams.

(j) Create logical data structures from process state vectors.

(k) Derive entity relationship model from system specification if required for database specification.

REFERENCES

Monk, A. (ed.) 1985. *Fundamentals of Human Computer Interaction.* London: Academic Press.

Shneiderman, B. 1987. *Designing the User Interface.* Reading, MA: Addison Wesley.

6

JSD
IMPLEMENTATION
STAGE
Physical system specification

This chapter covers JSD implementation stage, in which the concurrent model of the system is transformed into a scheduled system executable on a sequential von Neumann machine.

6.1 TRANSITION TO IMPLEMENTATION

The SSD is a concurrent system specification which could be implemented directly on several microprocessors (e.g. Inmos transputers or separate micros). Theoretically each instance of a process type could have a dedicated microprocessor to run on, but this would be an extremely inefficient use of resources so some transformation of the logical design is necessary to conserve processor resources. A possible transformation is to run each process type on a dedicated microprocessor. Because JSD processes are long-running, each microprocessor will have to be left permanently switched on even though it may be inactive for most of the time. Clearly this too is a waste of resources. The answer to the problem is to schedule execution of processes so that they run only when required.

Scheduling is a central concept in JSD implementation, which transforms a concurrent, perpetually active system into a sequential system in which processes have intermittent bouts of activity that take place only when necessary. Before considering scheduling, the designer has to decide on the implementation strategy. The system requirements may be for a centralized system to be run on a single mainframe or a distributed system with a network of micros running different parts of the system. For instance, a point of sale system could have filter processes running on separate microprocessors, while the main part of the system which does the stock updates runs on a mainframe.

The first step in implementation is to decide on the distributed/ centralized strategy and partition the SSD into implementation units. Figure 6.1 illustrates such partitioning for a system with distributed data capture feeding a centralized system on a mainframe.

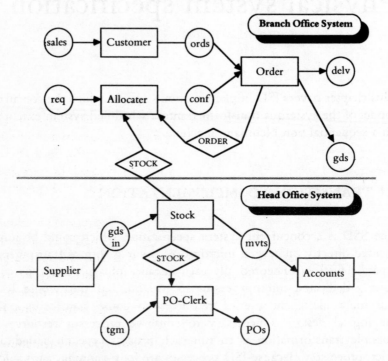

Figure 6.1 SSD showing implementation partitions.

When the allocation of processes to processors has been completed the next step is to choose the type of implementation. It may be that the processes need to share the processor on a time-slice basis, with each process running in turn for approximately the same period of time.

Operating systems may provide this facility in a multiprogrammed environment by time-slicing to ensure that execution time of each process on the single processor is shared equitably. However, either there may be some sequential dependency in the calling order of processes or each process may require a different amount of processor resource. Also, most computers at present require a system to execute as a single sequential process. These constraints mean that in most implementations a scheduler process is necessary to control the system's activity.

6.2 IMPLEMENTATION CONCEPTS

JSD has three key concepts in implementation:

■ Process inversion
■ Process scheduling
■ State vector separation

The fundamental transformation from a concurrent to a sequential system is achieved by *process inversion*. Inversion changes two communicating processes into a hierarchy of a subroutine and a main program. Interprocess communication is by parameters instead of messages. Instead of running perpetually, inverted programs run when required, the main program dictating the activity of its subordinate.

This concept can be used on sequential or parallel processing machines. For sequential architectures it can be used to change a sequence of interlinked processes into a chain of inverted programs and subroutines (see Figure 6.2). With concurrent architecture inversion may be used to schedule small groups of processes; in highly concurrent implementations each process is inverted to a simple scheduler which stores and retrieves state vectors.

Scheduler processes are added to the design during implementation to control the timing of execution and the calling sequence within the system. System processes are inverted with respect to the scheduler, which means that the scheduler becomes the main program and system processes become subroutines which are called by it. In the example system illustrated in Figure 6.3 below, the scheduler is introduced as a new process which calls other system processes.

The third concept, state vector separation, results from the JSD

(a)

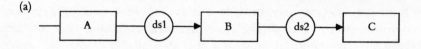

(b)

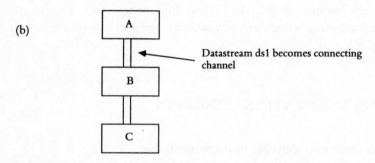

Datastream ds1 becomes connecting
channel

Figure 6.2 Inversion of a linear process sequence: (a) system specification diagram;
(b) inverted programs – system implementation diagram.

specification concept of multiple occurrences of a process type. Each
occurrence has an individual state vector associated with it. At implementation we have to arrange for physical storage of these state vectors,
which so far have notionally existed in the memory of a dedicated
microprocessor running each occurrence of the process type. State
vectors become separated from their processes as permanent records of
activity which are located in state vector files, more commonly known
as master files.

To illustrate state vector separation consider the case of the Account
process, in which each occurrence (Account-John, Account-Mary, etc.)
will have a master file record describing the status of John's, Mary's, etc.
account. The contents of these state vector records is specified by the
entity attributes discovered during analysis, with the addition of a key to
identify each state vector record uniquely. Thus the Account state
vector would probably have the composition:

Account number (key)
Account name
Account location (Bank sort code)
Current balance
Date of last update
Last update type (credit or debit)

JSD has advice to offer concerning data analysis (as described in Section 5.7) but conversion of logical data structures into a physical design for data storage and retrieval is not a concern of the method.

6.3 SYSTEM IMPLEMENTATION DIAGRAMS

System implementation diagrams (SIDs) show the calling sequence of processes in a hierarchical manner, with the scheduler at the top and inverted processes organized below in sequence. The calling sequence shows only possible lines of execution; to determine the conditions under which each process will be called the designer has to specify a PSD or structure text for the scheduler.

SIDs have four main components:

(a) *Process.* Shown as a rectangle. Generally there should be a one-to-one mapping of processes from the SSD to the SID, although processes may be subdivided in some circumstances (see the discussion of process dismembering in Section 6.9) and new processes will have been introduced for scheduling.

(b) *Inversion sequence.* Shown as a pair of parallel lines representing a datastream as a pipe. If more than one datastream connects two processes, two pipes are shown. The upper process calls the lower subordinate process which, in JSD terminology, is inverted with respect to its superior.

(c) *State vector file.* Shown by the normal flowchart file symbol. Processes that read SV files are connected by single lines to show the reading or writing relationship.

(d) *Buffer.* Datastreams are conceptually buffered queues. Inversion generally specifies zero buffering between processes; however, when two processes are connected and datastream messages are not passed on a one-for-one basis, a buffer file or working storage area will be required to store the waiting messages. Buffers are shown as ovals.

The various SID features are illustrated in Figure 6.3. Note that the calling sequence in the inverted chain illustrated in the diagram will be Deposit calls Interest which in turn calls Deposit Account. Each subroutine runs to completion and then calls the next until the bottom of the chain is reached, whereupon control is passed back up the line.

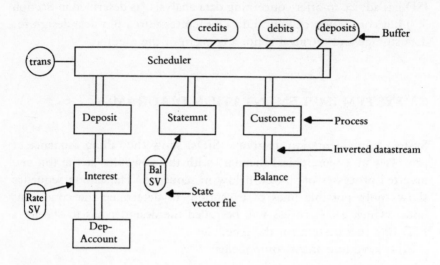

Figure 6.3 System implementation diagram: Banking system. Scheduler process is noted by a vertical bar.

6.4 SCHEDULER PROCESSES

Schedulers have specific responsibilities within a JSD system although the responsibility may be shared with other processes:

- The scheduling of system processes.
- Accepting system input and calling the appropriate process for that input.
- Buffer management of data being passed between processes.
- State vector management and updates.

These four tasks are usually the responsibility of schedulers; however, the latter three tasks may be shared or taken over by other system processes. For instance, data input and validation is handled by filter processes; and state vectors can be updated by the processes that own them. Where these tasks are integrally linked with scheduling, however, the scheduler process should be responsible. Thus when datastreams have to be buffered while being passed between two different inverted processing sequences, the buffer is managed by the scheduler which fetches the appropriate record from the buffer and passes it to the process that requires it.

All scheduler processes have a PSD (and if preferred a structure text) which shows how they will execute. New instructions are added to for implementation-dependent operations. The instructions are Call Process(ID) used to show the calls to inverted processes, Loadsv(SV-ID) when the scheduler accesses a state vector record from file and passes it to a process, Storesv(SV-ID) for the reverse operation, Read Buffer(ID) into datastream and Write datastream to Buffer(ID) for buffer management. A sample scheduler structure text is as follows:

Scheduler Structure Text

```
Scheduler Seq
    Read(Reply);
    Menu-Body Iter(while Reply not exit)
        Menu-Sel(Reply = 1)
            Deposit-Action;
            Call Deposit(FUNDS);
        Menu-Alt(Reply = 2)
            Report-Action;
            Loadsv(BAL);
            Call Bal-Report;
        Menu-Alt(Reply = 3)
            Update-Deposit;
            Read Buffer(DEPS) into DEP;
            Call Balance(DEP);
        Menu-Alt(Reply not 1-3,0)
            Display Err-Msg;
        Menu End
        Read(Reply);
    Menu-Body End
Scheduler End
```

In the above scheduler input data is handled by filter processes, apart from top-level system commands given by the user.

If the scheduler does accept all system input it should employ a rough merge, because the order of arrival of data from the outside world is often unpredictable.

6.5 PROCESS INVERSION

Inversion is a Jackson concept inherited from JSP. It changes the link between two processes from a file- or buffer-based connection into a direct call between the processes in which data are passed as parameters. The effect is that the decision about which process executes next is taken at the design stage rather than at run time. Control of process execution can happen in several different ways. The following scenarios are possible methods for controlling the execution of two connected processes:

■ Process A runs to completion writing records to the ds datastream buffer; when A has finished B starts and processes the records in the buffer. The ds buffer may be either unbounded, in which case a potentially unlimited amount of file store must be allocated to ds, or bounded to a set limit, in which case A runs until it fills the buffer, B empties it and A runs again and so on. This is not a true inversion.
■ Process A runs, produces a single ds record and then calls B which processes the record. On completion B calls A again. This is an example of inversion in which B is inverted with respect to A.
■ In the opposite case, B runs first until it encounters a Read(ds) instruction in its code; it then calls A to produce the record, consumes the record and the cycle recommences. In this case A is inverted with respect to B.

The two inversion types are shown in Figure 6.4.

Conversion of two concurrently linked processes to an inverted pair of processes is quite simple:

■ In the calling process, replace the Write(ds) instructions with Call Invert Proc Using(ds).
■ In the called (inverted) processes, replace Read(ds) instructions with Exit Invert Proc.

In simple inversions the main program and subroutine execute all their code each time they are called, but in most circumstances different parts of the code will have to be executed at different times. This is a consequence of the JSD specification of a long-running process. The model describes a complete life history; however, at run time only part

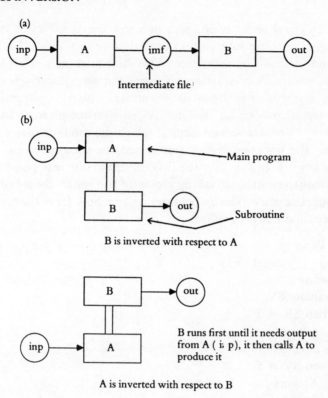

Figure 6.4 Variations of the inversion theme: (a) system specification diagram; (b) system implementation diagram.

of a life history will have to be executed to deal with the system input. To control partial execution of a process text JSD uses the text pointer.

Text pointers are required to implement most JSD processes. The pointer which is held in the process instance's state vector tells the process which part of its code should be executed to deal with the input record it has been given. In the book example, the text pointer might show that for *War and Peace* the next appropriate action is Return or Renew because the book status is 'on loan', while the pointer for Roget's *Thesaurus* indicates that Classify is the next action as the status indicated by the state vector is 'acquired but not classified' (see Figure 5.10). The text pointer tells the process where to start executing; the process text is then executed until the next halt point, whereupon the value of the state vector is updated and process execution halts.

Another example in which text pointers are necessary is in a data entry sequence. User interaction may halt at several points because the user wishes to finish an interaction. Also the system may halt interactive processes which are consuming resources but doing nothing when the user has walked away from the terminal. This sort of problem is handled by teleprocessing monitor systems on mainframes and mini-computers. There are several steps at which the monitor may swap and deactivate the program; when reactivated the program must resume where it left off. In this case the inversion uses the text pointer of the state vectors to control the calling sequence and restart the subroutine in the appropriate place. The text pointer records how far in the subroutine code execution has progressed.

```
Main Prog
    Call Sub using (SV)
Subroutine
    Evaluate SV
    When SV = 1
        Actions
        Save SV
    When SV = 2
        Actions
        Save SV
    When SV = 3
        Actions
        Save SV
End
```

The subroutine code is re-entrant, which means that it may resume at several points within the code. Each time a sequence is completed the state vector is updated and passed to the main program, when execution is halted.

6.6 MORE COMPLEX TRANSFORMATIONS

In theory conversion of a SSD to a SID could be a simple matter of following the connections on the SSD and building up a hierarchy of inverted processes. In practice the conversion is usually more complex.

Unfortunately systems are rarely as simple as two connected programs.

(a)

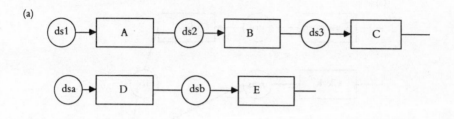

(b)

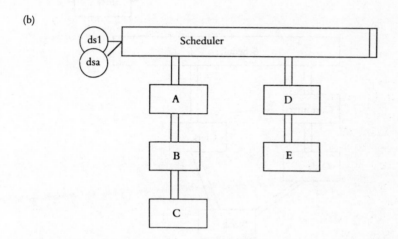

Figure 6.5 Simple implementation of SSD sequences: (a) system specification diagram; (b) system implementation diagram.

In a system with a linear sequence of connected processes transformation is simple (see Figure 6.5).

This can be inverted in a simple sequence of A calling B which calls C. But most systems are networks and pose problems of choosing which sequences to invert. System networks vary in their complexity, depending on the connections between processes. After linear sequences, tree networks are next in the scale of complexity. In tree networks there is only one pathway between any two processes, as shown in Figure 6.6.

In this situation implementation is still straightforward. In Figure 6.6 there are two inputs and therefore two inverted sequences are employed to process the input, the two lines meeting where the rough merge occurs. However, there may be two or more independent pathways to the same system process, creating a network like the one shown in Figure 6.7.

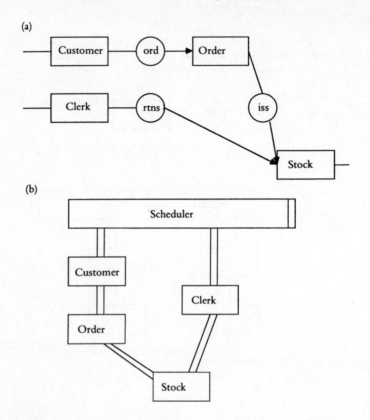

Figure 6.6 Implementing networks using rough merges: (a) system specification diagram; (b) system implementation diagram.

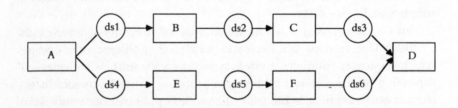

Figure 6.7 An SSD network with parallel pathways.

Here process D could be reached by either the BC or the EF route, and clearly we will have to decide which to use. The scheduler could devolve this decision to process A, which would then call either B or E depending on which type of record had been produced. To complicate matters there is no guarantee that either one or the other of the parallel pathways will be required; in some cases both pathways may have to be activated. To schedule the 'both' condition the SSD network would have to be dismembered, with the scheduler calling process A with BCD, EFD as separate inversion hierarchies. The scheduler would have to buffer data and call each pathway in turn (see Figure 6.8).

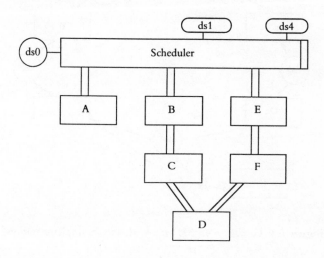

Figure 6.8 System implementation diagram for Figure 6.7.

Some systems are even more complex and contain circular pathways or circuits, as illustrated in Figures 6.9 and 6.10.

In the SSD depicted in Figure 6.9, the data passes from A to B and C but from C it may then go back via D and E to A. Obviously there must be an exit point (C to F), otherwise the system would produce no output. This type of system poses a scheduling problem because we have to ensure that downstream processes cannot 'read ahead' of upstream ones, i.e. that they cannot consume a datastream record before it has been produced. In an ordinary network this constraint might delay a process while it waits for the next record; in a circuit the problem is more serious because processes can lock each other in a 'deadly embrace' or deadlock. If A and B are each waiting to read a record from the other and cannot write one until they have done so, the system will come to a halt from which there is no escape.

This deadlock is inherent in the specification, and the only solution is to initialize the circuit with a dummy record to start the interlinked set

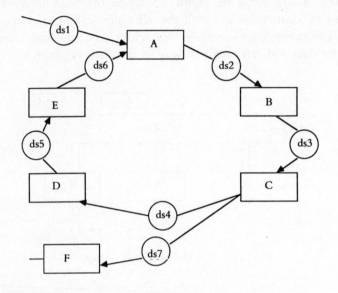

Figure 6.9 Circuit in an SSD network and its implementation.

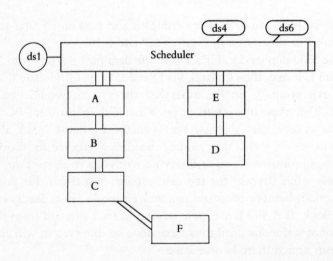

Figure 6.10 System implementation diagram for the circuit network in Figure 6.9

of processes cycling. In the example shown a dummy ds6 message would have to be provided when the cycle of the circuit was started. Implementation of the circuit could be a simple inversion of A through to F with buffers to store the output and input to the cycle (Figure 6.10). In implementation buffers have to be designed to preserve messages which are essential for the continuation process operation, otherwise deadlock may still occur.

Where the circuit has more inputs, such as a message-passing system with input arriving into any of the processes A to F, scheduling becomes more complex. The scheduler has to be able to call any of the processes A to F independently. Care has to be taken in the scheduler design to ensure that the Read–Write relationships of the processes do not allow deadlock to happen. To deal with cycles of activity the scheduler has to become a monitor which checks up on processes and makes sure they are in the right state before calling them. To do so schedulers use an 'inspect process state' operation to manage cycles, suspending and resuming process activity according to the requirements of the input and the state of the system.

Free-running processes

In some concurrent systems scheduling is not required and processes are run continuously on a single processor. Such processes are said to be *free running* in JSD. Free-running processes model entities in the real world which do not produce an event stream but instead have a state vector which can be sampled to determine what is happening. Switches and sensors are typical free-running processes. The implementation of these processes needs to be carefully designed to ensure that there are sufficient processor resources for correct execution, as otherwise another kind of deadlock, implementation deadlock, can be caused.

In JSD free-running processes often do not produce output datastreams; instead they just respond to events in the outside world and update their state vector. Examples are mechanisms such as switches, buttons and receptor devices which change state and then have their state inspected by the rest of the system. At implementation we have to allocate sufficient resources to ensure that the process can respond quickly enough to changes in the real world. To illustrate the point: if in a lift system there is a Button process which monitors the press on the lift call button, it has to run fast enough to record all the button press events. Hence if a button press action can happen in the minimum time span of 50 milliseconds, then the process must have a cycle time between two Get(SV) instructions of 50 milliseconds or less.

In some cases we may need to implement free-running processes in a multiprogramming environment which mimics separate processors. The processes are run on a single processor using time-slicing to mimic concurrency (see Figure 6.11). Time-slicing is usually provided by the hardware, and essentially divides the processor's attention between competing programs by giving each program in turn an equal slice of time. In this situation we need to ensure that the operation time allocated to each of our free-running processes is greater than the shortest time it takes for one state change.

Most time-slicing architectures are sophisticated enough to divide the time up between processes according to their needs, with inactive processes receiving less time than active ones. Schedulers can be devised for the same effect in real time systems in which process execution time becomes vital.

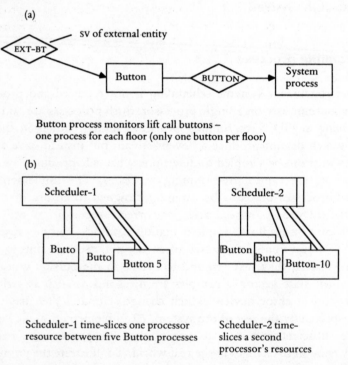

Figure 6.11 Free-running processes: (a) system specification diagram; (b) system implementation diagram. Note that further processors would be required to implement other system processes.

6.7 CHANNEL SCHEDULING

During implementation the developer has a choice of the pattern of inversion to follow. A strictly hierarchical approach may be chosen but this does create inflexibility in the system, e.g. the calling sequence is predetermined so that A always calls B which calls C, etc. The alternative approach is to adopt a 'flat' implementation with all processes called directly from the scheduler. This is called *channel scheduling*. The advantage of channel scheduling lies in its flexibility. The scheduler can determine when each process should run according to the input and processes' status as recorded in state vectors. Many interactive systems require this flexibility to run processes independently according to user demand. Hierarchical scheduling is more difficult with rough merges and loops in networks, and it also requires more complicated buffer management.

Channel scheduling may also be necessary in networks in which processes execute in an unpredictable manner. Consider a processing sequence with three processes A, B and C as illustrated in Figure 6.12. Process B may run for a time and if it completes its activity passes a message, ds2, to C; but it may not complete its activity because of competition from other processes in a multiprogrammed machine. Process B will have halted at a predefined point, either a Read(ds1) or a Write(ds2), but the scheduler needs to know at which point. This problem is similar to the one of the teleprocessing monitor, which we solved with state vector control of inversion. In this case the scheduler has the task of monitoring the progress of system processes.

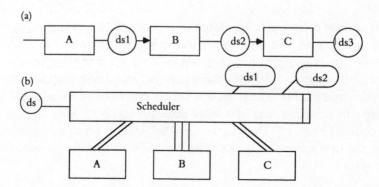

Figure 6.12 Channel scheduling: (a) system specification diagram; (b) system implementation diagram.

The scheduler has to inspect the status of processes under its control to determine which should be called next and when. To do this a new instruction, Query Process(ID), is used which inspects the process's state vector and returns a message Active/Inactive. By polling each process in the channel in turn the scheduler can monitor activity and call the appropriate process in the sequence when activity resumes. The structure text for a channel scheduler might be:

```
Channel Scheduler Iter
    Scheduler-Body Seq
        Query B;
        Decide Sel(Bsv shows Read)
            First-Part Seq
                Call A(ds0);
                Call B(ds1,ds2);
            First-Part End
        Decide Alt(Bsv shows Write)
            Second-Part Seq
                Call B(ds1,ds2);
                Call C(ds2);
            Second-Part End
        Decide End
    Scheduler-Body End
Channel Scheduler End
```

6.8 TRANSFORMATION GUIDELINES

This section offers some heuristics to follow which tackle the questions of where to start and how to proceed when scheduling.

Apart from complexities imposed on scheduling by the form of the network, in on-line systems there may be constraints imposed by the users' requirements which dictate which processes must be capable of being run independently. To solve these problems we need to segment more complicated SSDs into separate scheduling groups of processes. The following steps may be applied to SSDs to specify the scheduling groups:

(a) Examine user interface/task requirements to determine which processes must execute separately. This can also be effected by following inputs (both data and user commands) and noting which processes they will invoke. This analysis will produce 'cuts' in the SSD connections and subdivide the SSD into separate inversion pathways. Cuts will generally fall on state vector connections, as state vector inspects frequently belong to information processes which are run independently of main processes.

(b) Within each pathway (or the whole system if no separation has been necessary), identify system inputs.

(c) Invert the first process which accepts the input (this will probably be a filter process) with respect to the scheduler.

(d) Continue inverting the first process's nearest neighbour, and so on until the output is reached.

(e) When cross-links are found, examine the user requirements to see if each pathway has to be called independently. If so, separate the two

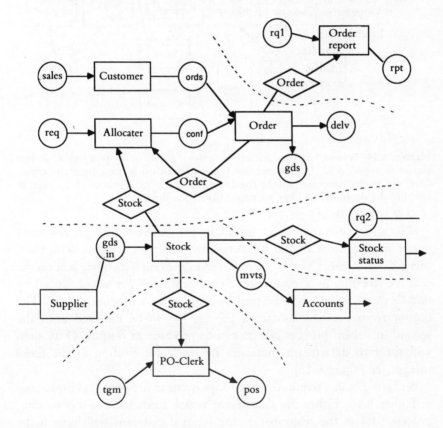

Figure 6.13 SSD of an order-processing system, showing cuts in the network.

pathways and connect both to the scheduler; otherwise connect the first input process on the other pathway to the scheduler and complete the inversion hierarchy.

The sample SSD showing cuts in the network resulting from user requirements and the SID derived from the SSD are shown in Figures 6.13 and 6.14 respectively.

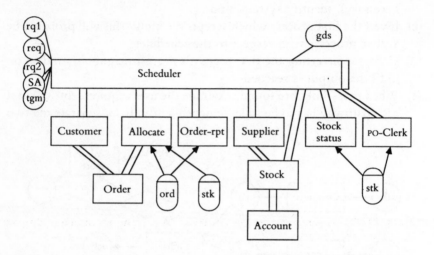

Figure 6.14 System implementation diagram of the order-processing system shown in Figure 6.13. Implementation follows datastreams from input to output. Only one buffer is required for the fixed merge into the Stock process. Note that, to simplify the diagram, no filter processes have been shown.

Whenever convergence is found in the network the provision of buffer management should be considered for fixed merges. With fixed merges datastreams have to be buffered to prevent data being lost on the inactive streams. In fixed merges datastream messages are produced by two or more processes to be read by a single downstream process. The output from the first upstream process has to be buffered until the second upstream process can be run to produce its output. Only then will the two datastream messages be ready for reading in the fixed merge (see Figure 6.15).

Buffering is also required when loops occur in networks, as illustrated in Figure 6.16. Either the datastream which feeds back to a preceding process (ds4 in the diagram) or the input datastream will have to be buffered.

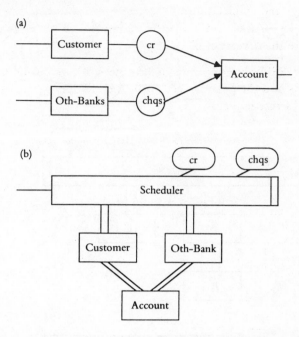

Figure 6.15 Implementing a fixed merge: (a) system specification diagram; (b) system implementation diagram.

6.9 IMPLEMENTING BATCH PROCESSES

Batch processes can be implemented directly from model processes with a scheduler which may be no more than a few JCL cards to invoke the job. Sometimes batch programs can present a problem at implementation time when the user requirements for running parts of the program clash with the original model.

JSD processes are modelled as long-running, so they can take care of all the events in a process's life history. But some batch processes need to be run infrequently and represent only part of a processing history. It is inefficient to have a process running a whole life history which is going to be only partly used. To deal with this inefficiency JSD has a procedure called *process dismembering* which segments a process's life history into the parts necessary for implementation as separate batch programs.

Process dismembering may be used for other reasons, such as

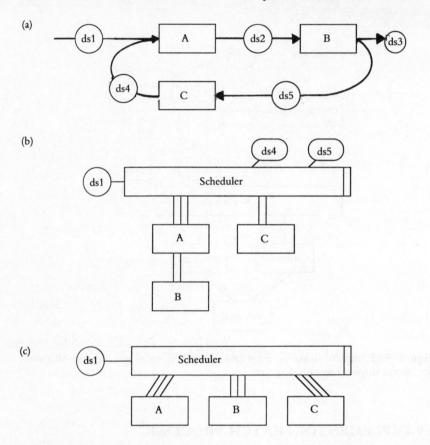

Figure 6.16 Implementing loops in networks: (a) system specification diagram; (b) system implementation diagram. The Scheduler stores output from C (ds4), waits for the next input ds1 message, then calls A and B, passing them ds1 and ds4. Process C could either be called directly after A and B or run separately if the rates of production of ds1 and ds4 are different. Alternatively channel scheduling could be used, as illustrated in (c).

optimization and physical machine constraints. For example, programs may need to be rearranged to fit into load units, and response time considerations may require optimization for speed of execution.

To illustrate batch process design, consider the Bank Deposit example. The original version had Credits and Debits arriving unpredictably throughout the day. The Bank of Ruritania wishes to buy the original version but only processes transactions once a day and it is the practice to presort the Credits and Debits. The process structure for the system is

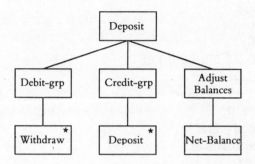

Figure 6.17 Process dismembering: Deposit system for the Bank of Ruritania.

shown in Figure 6.17.

So far we can convert our on-line process into a batch process with only minor changes. But then the Bank of Ruritania changes its mind and wants to process credits and debits separately – the debits at close of business and the credits overnight. In this case we have to dismember the original process because we require the credit and debit parts to run at different times. As a result we end up with two separate programs, CR-Deposit and DB-Deposit, but one action, Calculate Balance, which requires both. To get around this problem we use a common data area on file which can be accessed by both programs so a total can be calculated and printed by the last batch to run. The resulting structure text for the two programs is:

```
DB-Deposit Seq
    Open DB;
    Init counters;
    Read DB;
    Debit-Body Iter(until Eof)
        Total debit;
        Total transactions;
        Read DB;
    Debit-Body End
    Write TB;
    Close DB;
DB-Deposit End
```

```
CR-Deposit Seq
   Open CR;
   Init counters;
   Read CR;
   Credit-Body Iter(until Eof)
      Total credit;
      Total transactions;
      Read CR;
   Credit-Body End
   Calculate Balance;
   Write BAL;
   Close CR;
CR-Deposit End
```

Process dismembering has a special representation on the SID: the dismembered programs are uniquely identified by a prefix which is added to the name of their parent process (see Figure 6.18).

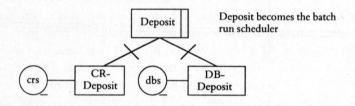

Figure 6.18 Deposit system: system implementation diagram. The Deposit process is implemented as two dismembered processes, CR-Deposit and DB-Deposit, which are called by their parent process, Deposit. Calling of batch programs is shown by the bar on the connecting line.

Batch processes will frequently require a scheduler process of their own to control execution timing and manage buffers, as datastreams are no longer processed on arrival but accumulate in an input file which is read at a specific point in time.

6.10 TRANSITION TO CODING

At this stage the JSD specification is nearly complete, and for experienced programmers it may suffice as a full program specification from which they can code. It is advisable, however, to add some additional detail in the form of elementary operations to make the specification more complete. At this stage physical design detail is added to the specification for opening and closing files, initializing counters, etc. To do so JSD borrows the latter steps of JSP, which conveniently fit on to the end of implementation as an elaboration phase converting the PSD and structure text into a program specification.

The three additional steps are:

- List elementary operations for the process.
- Assign elementary operations to components on the PSD.
- Assign detail to elementary operations and conditions to components.

Elementary operations consist of two things:

- Detailed specification of actions.
- Computer-related operations not yet specified.

Elementary operations are shown as small boxes which are attached to the component within which they will reside. Thus the first components in many PSD structures will have elementary operations dealing with initializing variables and opening files, the last component will have operations for closing files, and actions will have operations incrementing counters, assign statements, displays etc. Elementary operations are then assigned to program components. This requires the answer to two questions:

- Which component should the elementary operation be assigned to?
- If there is more than one operation per component, in which order should the elementary operations be executed?

The answer to the first question is generally given by specification in the earlier stages of development of operations for the logical parts of the process specification, but where to place physical details of design may not be so clear. The following guidelines can be used, although detailed design ultimately depends on the application in question:

(a) Find the operations for file opening and initialization. Assign these to the first component in a program structure.
(b) Find the input operations (file reads, input accepts). Assign these as the first elementary operation in the action component for which they provide data.

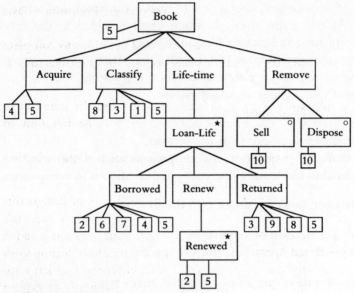

Elementary operations list:

1 Initialize loan counter   LOANCTR := 0
2 Increment loan counter   LOANCTR := LOANCTR + 1
3 Set Boolean In.library = True INLIB := TRUE
4 Set Inlibrary to false   INLIB := FALSE
5 Read Inrec   READ BOOKREC, (RECTYPE, BOOKNAME, READER-ID)
6 Assign loan details   TITLE := BOOKNAME, BORROWER :=
                                        READER-ID
7 Assign, loan date   LOANDATE := TODAYDATE
8 Initialize loan time   LOANDURN := 0
9 Calculate loan time   LOANDURN := RETURNDATE−LOANDATE
10 Delete Book   DELETE BOOKNAME (SV RECORD−BOOK-ID)

Figure 6.19   PSD showing elementary operations.

(c) Allocate computational and algorithmic operations to the appro-
     priate actions in the process.
(d) Find and allocate any status management operations, e.g. incre-
     menting counters, within the condition or iteration subtree to which
     they belong.
(e) Allocate termination and file-closing operations to the final com-
     ponent in a process structure.

The next step refines elementary operations by adding detail necessary
to translate operations directly into code. Detail is added to elementary

operations in pseudocode or in the syntax of the target language to specify procedural detail (e.g. Add 1 to Linecounter, Premium=Basic Cost*Weighting).

Conditional logic is added to the selections and iterations to complete the specification. When the conditions are specified the questions to bear in mind are:

- For iterations: what are the conditions which cause the iteration to terminate? Or what condition must be true for the iteration to continue?
- For selections: are the conditions for each branch of the selection exclusive? Has an 'else' condition been specified?

Although these steps have been placed in the final part of design this does not preclude their incorporation in earlier stages. It is quite natural, for instance, to add conditions whenever a Selection or Iteration is added to a PSD diagram. Additional computer-related operations dealing with file I/O and other physical machine matters are, however, best left until the later stages of implementation to prevent machine issues obscuring the logical part of the design.

Operations are referenced by a simple number on PSDs as shown in Figure 6.19. Operations are held in a list and actions and conditions are assigned against the list which, with the process structure diagram, forms the coding specification. Translation to code in a target language is then a direct matter of mapping program components on to constructs in the host language. While this is usually straightforward, implementation of backtracking and inversion require further consideration.

Implementation of backtracking

The Posit and Admit components are implemented as ordinary program structures such as Pascal Procedures or COBOL Performs. Quit statements are tests for the unexpected and are conditional jumps to the Admit branch and are implemented as Goto(Admit part) statements. The Goto statement has received a bad press in computer science literature for several years; its use in a controlled manner is helpful, however, because it makes the program text easier to understand by separating the normal from the abnormal processing and by making exit points from the normal sequence explicit. If backtracking was not used small procedures for remedial action would have to be embedded throughout the normal process text. The whole process becomes a deep-nested sequence of selections which makes the specification text less tractable.

Each Quit statement is a condition testing for the unexpected. Attached to the condition is a Goto statement which is invoked if the condition is true. The Goto then executes the Admit branch of the back-tracking program. Quits can be linked to different parts of the Admit branch using a text pointer, so processing of side effects can be linked to specific Quits. Another Goto is added to the end of the Posit part of the program to jump over the Admit branch in the event of none of the Quits being triggered. The use of Gotos is illustrated in the

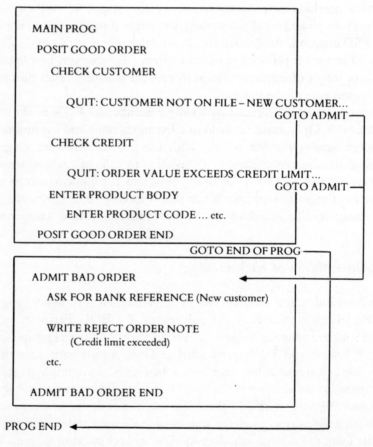

Figure 6.20 Backtracking: schematic view of program source code.

program schematic in Figure 6.20. The equivalent in COBOL–85 code
would be:

```
PROCEDURE DIVISION.
MAIN-PROGRAM.
   PERFORM INITIALIZATION
   PERFORM VALIDATE
   PERFORM TERMINATION
   STOP RUN.

VALIDATE.
   DISPLAY FORM-SCREEN
   PERFORM POSIT-BRANCH THRU BACKTR-END UNTIL EOFINPUT.

POSIT-BRANCH.
   ACCEPT CUST-ID
   IF CUST-ID NOT NUMERIC THEN
      MOVE 1 TO ERROR-CODE
      GO TO ADMIT-BRANCH
   END-IF
   ACCEPT PRODUCT-ID
   IF PRODUCT-ID > 999 OR PRODUCT-ID < 1 THEN
      MOVE 2 TO ERROR-CODE
      GO TO ADMIT-BRANCH
   END-IF
   *************etc

   GO TO BACKTR-END
   (* end of Posit *)
ADMIT-BRANCH
   EVALUATE ERROR-CODE
   WHEN 1
      DISPLAY"Numeric customer number required"
   WHEN 2
      DISPLAY"Product code should be within the range of 1 to 999"
   WHEN 3
   *************etc

   END-EVALUATE.
   (* end of Admit *)
BACKTR-END.
   DISPLAY CONTROL-MENU
   ACCEPT USER-REP
   EVALUATE USER-REP
   WHEN 0
      SET EOFINPUT TO TRUE
   WHEN 1
      PERFORM SAVE-REC
   WHEN 2
      PERFORM EDIT-REC
   END-EVALUATE.
(* End of backtracking prog *)
```

Backtracking can be implemented without Gotos by using a deep-
nested series of 'if-then-elses' but the resulting program text is more
difficult to understand.

Implementing inversion

Write and Read statements in the concurrent process specification are changed into Call subprogram and Exit statements respectively. This change can be illustrated with the library book structure text:*

```
Book Seq
    Acquire Seq
        Read(Inrec) ──────────────────────────►exit program
        Inlibrary := false
    Acquire End
    Classify Seq
        Read(Inrec) ──────────────────────────►exit program
    Classify End
    Book-Life Iter(while loans)
        Loan-Body Seq
            Loan Seq
                Borrower := Reader-ID
                Lenddate := Date
                Inlibrary := false
                Loancntr := Loancntr+1
                Read(inrec) ──────────────────►exit program
            Loan End
            Renew Iter(while Loancntr<4)
                Renewed Seq
                    Loancntr := Loancntr+1
                    Read(Inrec) ──────────────►exit program
                Renewed End
            Renew End
            Return Seq
                Inlibrary := true
                Loandurn := Date−Loandate
                LoandurSV := Loandurn
                Loandurn := 0
                Read(Inrec) ──────────────────►exit program
            Return End
        Loan-Body End
    Book-Life End
```

* := is used for 'is assigned to'.

```
      Remove Sel(Inrec = Sell-type)
        Sell Seq
          Delete BookSV(Book-ID)
          Read(Inrec)─────────────────────────────►exit program
        Sell End
      Remove Alt(Inrec = Dispose-type)
        Dispose Seq
          Delete BookSV(Book-ID)
          Read(Inrec)─────────────────────────────►exit program
        Dispose End
      Remove End
  Book End
```

While the calling program text will not require much further change the subprogram text probably will. If a process which is inverted has many Read statements then it will also have many exit points. This is liable to be true for most model processes. The problem is how to control the entry points into the subroutine text so that only the relevant piece of code is executed each time. The solution to this problem lies with the role of the text pointer.

The text pointer is passed down to the inverted program as it is called. The value in the pointer is used to determine which piece of the subprogram text to execute. This could be coded using Gotos to transfer control as follows:

```
Goto ST(value); ST1.
Book Seq
    Acquire Seq
        Inlibrary := false
        ST := 2; Return; ST2.
    Acquire End
    Classify Seq
        Inlibrary := true
        Loancntr := 0
        Loandurn := 0
        ST := 3; Return; ST3.
    Classify End
    Book-Life Iter(while loans)
        Loan-Body Seq
            Loan Seq
                Borrower := Reader-ID
```

```
            Lenddate := Date
            Inlibrary := false
            Loancntr := Loancntr+1
            ST := 4; Return; ST4.
      Loan End
      Renew Iter(while Loancntr<4)
            Renewed Seq
                  Loancntr := Loancntr+1
                  ST := 5; Return; ST5.
            Renewed End
      Renew End
etc. . . .
```

The main program calls the subroutine, passing it an input record and the text pointer. The main program has to evaluate the input record and access the appropriate state vector, in this example using the Book-ID key. The text pointer within the state vector will be set to point at the next part of that process instance's life history (e.g. Roget's *Thesaurus* has just been classified, therefore it expects the next action to be a loan with a text pointer set to ST3). Control is switched to the appropriate part of the subroutine text by evaluating the text pointer value in the Goto statement. The subroutine runs until the next Return is encountered. Just before the Return the value of the text pointer is updated to point at the next piece of the program's life history. Note that once the subprogram text has run the state vector is updated before control is returned to the main program. Inversion control can also be implemented using Case statements in Pascal or its COBOL-85 equivalent, Evaluate, although this necessitates some design of the program code into paragraphs which correspond to the text between two Gotos. The paragraphs can then be called from the Case statement by evaluating the text pointer value.

6.11 SUMMARY OF IMPLEMENTATION

Implementation changes the concurrent system model of the SSD into a physical design which can execute on a sequential machine. This transformation is achieved by introducing a scheduler process and inverting other system processes as subroutines which are called by the scheduler.

Inversion changes processes which communicate by datastreams into a main-program – subroutine relationship which pass data as parameters. Sequences in the SSD become calling hierarchies in the SID. Various heuristics can be employed to refine the design during implementation and to separate a complex SSD network into inversion hierarchies.

State vectors are separated from their processes during implementation and placed in master files, each state vector becoming a record with its unique identity preserved in a key. Datastreams may become single record parameters or buffers if an accumulation of messages is anticipated. Management of buffers, input data, state vectors and calling processes is generally the responsibility of the scheduler.

Batch programs may have to be implemented by separating the pieces of a process which need to be run independently at different times. The dismembered processes are then scheduled separately.

When the implementation transformation is complete, further details may be added to PSDs to complete the program specifications.

Key points

- Implementation converts the concurrent system specification into a sequentially executable system.
- Implementation is carried out by introducing a scheduler to control process execution and by inversion.
- Inversion converts two processes connected by a datastream into a main-process – subroutine hierarchy.
- Channel scheduling can be used to make the implementation more flexible and reduce the need for buffers.
- When fixed merges are present one of the datastreams will have to be buffered.
- State vectors are separated from their individual processes to become state vector records in files.
- State vectors and the text pointer are used to control the execution of inverted processes.
- Free-running processes need no scheduling but must be allocated to a processor which allows them to run fast enough to respond to all events.
- Processes may be dismembered for batch processing if different parts of their text are to be run at different times.

Key steps

(a) Divide SSD into scheduling groups according to system requirements of distributed processing, hardware and user interface needs.

(b) Add a scheduler process for each scheduling group.
(c) Start with system inputs and invert input processes as subprocesses of the scheduler.
(d) Following datastream connection in the SSD, invert nearest neighbours to input processes as the next subprocess level and continue until output is reached.
(e) Add buffers for fixed merges and loops in the network.
(f) Add state vector files and show access paths to processes.
(g) Convert process specifications to inverted form by changing Writes/ Reads for Calls and Exit-programs.
(h) Add design detail by listing elementary operations and conditions.
(i) Assign elementary operations and conditions to program components.
(j) Add further detail to elementary operations and conditions in pseudocode or as statements in the target language.

7

MANAGING JSD DEVELOPMENT

This chapter deals with a collection of issues which are not a central part of the method. Issues of introducing and managing JSD projects are reviewed, together with prototyping approaches and questions pertaining to user–analyst communication.

7.1 INTRODUCING JSD

JSD can be perceived by analysts unfamiliar with structured techniques as an alien method which is hard to understand. This reaction may come about either because of a wide knowledge gap between structured methods and more classic techniques, or as a reaction to formalism in design and specification. JSD is a concise method which creates a detailed specification during the first few steps and this can cause difficulties, particularly in the early stages of analysis. Whatever the reason, such reactions may endanger attempts to introduce JSD into an organization so it is worth devoting a little space to the subject of starting out with JSD in a new environment.

To combat difficulties of introducing JSD it is necessary to discover which of the method's complexities are difficult to comprehend. Unfortunately in JSD the complexity and comprehension difficulties come in the same place – during requirements analysis. As this is the bit most people try first, it is not suprising that they may be put off the method after an initial encounter, especially if it is not clearly explained.

Once the first effort has been made, most people find the method becomes progressively easier to use and find its formalism appealing.

While some of these problems can be alleviated there is no complete answer to the difficulties of comprehension. In the end anything worth having has to be worked for, and JSD is no exception. Difficulties can be anticipated, however, allowing tutors to draw attention to problem points. The problems which may arise on first acquaintance with JSD may be summarized:

■ Being locked into functional thinking.
■ Not knowing where to start finding entities.
■ Bewilderment induced by the lack of guidelines in the early stages.
■ Difficulty in grasping the concept of time ordering.

The last problem is the easiest to solve. Most people understand PSDs after examining a few worked examples, but the other problems are more serious. Thinking functionally is a common complaint which distracts analysts from creating a good JSD model. Only experience with analysis of time ordering and modelling will make a good JSD analyst, although some help can be given by examples of the differences between functional and JSD system models and by encouraging a questioning and critical attitude to the system description. JSD modelling is essentially creating an accurate description of the system resulting from questions about all the permutations an entity's life history can undertake. Good analysts have always done this; JSD provides a framework which encourages all analysts to model systems activity correctly.

The lack of guidelines and getting started are linked problems. The best advice is to start with the central problem of the system and define the objects which are associated with it. The main point to remember is to think 'object' while trying to find entities. It is important to try an initial analysis of entities and then rework the ideas, as the first few attempts may well be wrong. JSD is an iterative process which produces a detailed understanding of the system in its early stages; part of the price of detail is getting the wrong initial view and having to start over again.

The other point to bear in mind when introducing JSD is that although it is a fairly comprehensive method it does not cover everything. Its omissions need to be covered with other methods and techniques.

The benefits of using JSD will be more accurate and reliable system specifications, and system designs that are easier to maintain. These benefits will take time to become apparent. On introduction of JSD the balance sheet is negative; costs have to be incurred in training and the efficiency of development staff in using the method will not be optimal while they progress along the learning curve. But once JSD systems begin to be implemented the benefits should begin to accrue as fewer errors appear in systems, fewer modification requests are generated by poor requirements analysis, and the necessary modifications are implemented more easily.

7.2 JSD SPECIFICATION: OMISSIONS AND ADDITIONS

The major omission in early versions of JSD was specification of data structures and relationships. Since then JSD has been elaborated to cover data analysis, and now produces specification of data structures and relationships. It cannot, however, deal with data analysis for conversion to a relational database, for which full normalization is required, but this deficiency will not trouble analysts on the majority of projects.

JSD, in common with other structured methods, omits some parts of a full system specification. Screen and dialogue design will have to be added although JSD does have a specific part of the system (filter processes) in which the man–machine interface should be placed. Computer-related activities are deliberately ignored in modelling, which aims to analyse only the logical elements of a system; hence system security and backup have to be added to the design in the function step. Recovery, restart and security procedures are similar physical design considerations which should be added as interacting functions in the JSD scheme.

JSD does not cover cost benefit analysis, estimates and scheduling; these are project management and business analysis issues which can be applied to any structured development method. JSD helps these tasks because its discrete stages make estimating and scheduling more rigorous. With these 'extras' JSD can be applied to a wide variety of systems.

7.3 JSD PROJECT MANAGEMENT

There is nothing explicit in JSD which specifically addresses project management issues, but features of the method do lend support to project control. The issues which any development method should address may be summarized as follows:

- Clear phases in development bounded by stop and start points.
- Deliverables from each phase.
- An organizational framework for the whole development.
- Ability to subdivide and integrate large projects.

Before each project phase begins, management must estimate and schedule work for the phase, in order to plan resources and work loading. Once the project is under way check points are necessary to monitor any slippage behind schedule, along with quality control of deliverables to check orderly progress.

JSD organizes development into phases which have clear end points and each phase has a set of deliverables. This enables systems to be built up in an orderly sequence; the initial model is constructed first, followed by user requirements and then the user interface. These phases also permit flexibility in development because different parts of the system may be analysed and specified in turn; for instance, only part of the whole system may be modelled and functions added to this subsystem before continuing with modelling other parts of the system. This creates a horizontal flexibility for management to choose the point at which each part of the system should be developed and thereby to control the workload. The JSD workpackages in a complete system implementation are shown below.

JSD project phases in different parts of a system

Initial analysis	Inputs +outputs	Timing+ General implementation	Detailed implementation
JSD entities	Filters	Timing	Coding
Model processes	Functions	Strategy	Testing
	Interacting functions	On–line distributed	Data load Parallel run
			Cut over

Projects usually come with time and cost constraints; in addition, systems may have differing degrees of precision in the specification of user requirements which make demands on the adaptability of a method. Consider two sets of contrasting system constraints, one set for a quick-decision support-type system, the other for a significantly large system:

Low cost	All budgets considered
Rapid development needed	Time scale in long-range plan
User requirements vague	Well-defined activities and requirements

The project controller has to answer questions concerning which part of the system should be developed, whether the approach should employ full analysis, specification and development in COBOL, PL/1 etc., or whether prototyping and a fourth-generation language should be used.

JSD can fit into many different development scenarios. For classic developments in a standard language it has a well-ordered framework of development stages, but its horizontal flexibility allows several alternative solutions to be offered to the user. The whole system can be modelled and then different automation boundaries can be drawn on the SSD to give the user alternative designs with different cost implications. One tactic is to identify a number of desired outputs and trace their connections back to model processes. This gives a clear picture of the number and type of processes which must be implemented for each solution. Cost estimates can be made for each solution based on the number of processes which have to be implemented.

At first sight it may seem JSD is not appropriate for prototype development with fourth-generation languages. This approach is likely to be employed on systems with vague user requirements which have information analysis and reporting needs. With a little reorganization, however, JSD can fit into a prototyping life cycle. With prototyping it is expected that the model will need refining in the light of experience so specification becomes iterative. The JSD method of modelling should help to clarify the user's and the analyst's perception of the system as it evolves. A possible prototyping cycle could be as follows:

Initial analysis	Build prototype	Test	Rebuild prototype	Operational cut over
JSD model	Add functions schedule		Redefine model	
	Implement processes with 4GL	Trial	Reconstruct some processes	Data load Full test cut over

JSD has the flexibility for phased development and can be used in different development contexts, from the classic DP department approach to prototyping. The emphasis on separating model processes from functions allows addition to the system without major redesign and the ability to develop a complete system as a series of subsystems provides for flexible management.

7.4 DEVELOPMENT TOOLS

Structured development methods create large volumes of diagrammatic documentation which is difficult to maintain by hand. JSD is no exception in this regard. To reap the full benefits of productivity increases, automation of documentation is necessary with computer-based diagram drawers, specification text editors and data dictionaries.

JSD specifications can be documented using Speedbuilder, which holds specification details in a data dictionary and performs consistency checking. Project development is supported by JSD-FRAME, which controls development phases and provides the project management support for critical path, estimating budgeting etc. Process structure diagrams can be designed and documented with PDF (Program Development Facility).* Although support tools undoubtedly help structured methods by making analysts and designers more productive, extra gains in productivity are possible if code can be automatically generated from specifications. System development and maintenance then become a matter of designing and changing specifications rather than code. JSP-COBOL allows the production of COBOL code from PDF specifications with minimal addition of code for details of elementary operations.

JSD is fully supported by development tools and automated code

---

* PDF is the property of the UK Atomic Energy Authority.

generation; the reader is referred to Michael Jackson Systems Limited, London, who will be able to supply further details of the products.

7.5 COMMUNICATING WITH USERS

A development method should be tractable so that users can understand and criticize it, and help the analyst to construct specifications. JSD is quite formal in its notation and creates a detailed view early on in analysis. These properties do not help user–analyst communication, however, because the diagrams are not easily understood by novice users, which poses a problem when using JSD in an end-user environment. One solution is to keep two sets of documentation, a full version for the DP department and a simplified set for the users.

Unfortunately JSD documentation is not easy to simplify. Diagrams are easier to understand than structure text, but PSDs and SSDs cannot be significantly simplified without losing accuracy. The choice is whether to expose users to these diagrams and run the risk of confusing them, or alternatively to maintain a separate set of nontechnical user documentation constructed solely of narrative.

There are arguments to support both approaches. Translation into narrative involves the analyst in more effort and, because of the nature of natural language, introduces ambiguity into the specification. To use technical documentation means that considerable effort has to be made to teach users about PSDs and SSDs, although diagrams can form a powerful method of communication. PSDs in particular present a clear picture of processing sequences which the analyst and user can 'walk through' to make sure the analyst's perception is correct. Once familiar with JSD diagrams, users quickly spot errors and inconsistencies.

7.6 SUMMARY OF MANAGEMENT ISSUES

JSD requires considerable initial investment in training for analysts, because its view of systems development is formal and different from other methods. Once that investment has been made, difficulties may be experienced with the early stages of JSD analysis in which the analyst gets little guidance. It may take some time for analysts to 'think Jackson' but once they do so, the benefits from more accurate system specification and better system maintenance should more than repay the initial investment in training and any extra effort expended on specification.

JSD specifications omit dialogue design and user interface specification, although special processes (filters) are identified as the location for the system interface. Other omissions are computer-related procedures such as file maintenance, backup, recovery and restart, although these may be added as interacting functions. It may be necessary to employ a data analysis method and normalization to supplement JSD if a database conversion is being undertaken. JSD, in common with other structured methods, does not cover issues of costing, estimating and scheduling of system development.

Project control is helped by the organization of JSD into clear developmental steps which have deliverables associated with each stage. JSD provides a clear definition of progress throughout development and allows flexibility of project control by partitioning the system into subareas which may develop separately. This enables alternative solutions to be presented to users. JSD can be used either within the classic system development life cycle or, because of its incremental approach to specification, in prototyping.

JSD documentation may present problems to end-users unfamiliar with the diagram conventions. Either documentation can be translated into narrative for users, or it may be worth while educating users to read JSD diagrams which can then form a powerful vehicle for analyst–user communication.

Key points

- Introducing JSD should be undertaken only with support and training.
- The most common difficulties are 'functional thinking' and getting started with entity analysis.
- JSD delivers more reliable maintainable systems by clearer specification and correct problem analysis.
- The method has clear phases which enable flexible but strict project management.
- JSD can be used in a variety of ways from complete system developments with third generation languages to prototyping.
- The method is supported by tools for developing specifications and automatic generation of COBOL code.

APPENDIX A

CASE STUDY 1
Bank deposit forecasting system

A.1 BACKGROUND

The First National Bank of Ruritania (FNBR) is interested in attracting money from very rich clients. As nearly every other bank has the same interest, the FNBR offers its customers special rates if they place large sums of money on deposit for a set number of days, typically 30, 60 or 90 days. The FNBR then pools the money from several customers and places it on deposit with other banks at a higher interest rate than is obtainable for smaller sums. The FNBR makes money from the difference between the interest rate it pays to its customers and the rate it receives from other banks.

Unfortunately some of the FNBR's customers have their own views about where their money should be placed and maintain a blacklist of banks which they consider not creditworthy. The FNBR undertakes, in the interest of good customer relations, not to place customers' money with banks on their personal blacklists. Therefore the art of successful money dealing becomes a question of placing the maximum amount on deposit in the bank with the most favourable interest rate within the constraints of individual customers.

The manager of deposit trading operations has asked us to investigate automation of the system to support this activity, known to the bankers as *fixed-term deposit dealing*.

Initial analysis: deposit recording

The terms of reference are to provide automated support for the deposit dealing and recording operations. The physical transfer of money is not within our brief. The following narrative is a résumé of the analyst's first interview with personnel in the user department.

Two groups of people run this system: the dealers who decide where to place the money and the order clerks who make sure the paper work is carried out to record the deposit. From an interview with the chief order clerk we learn the details of the procedure.

When a dealer concludes a deal with another bank, he fills out a ticket which records the money deposited by customers, the currency value, rate of interest and the receiving bank. The ticket is passed to an order clerk who checks the bank and customer IDs against a list, sense-checks the ticket and then raises a query with the dealer if necessary. The dealer amends the ticket as appropriate. When the order clerk is satisfied he partitions out individual deposits by customer, calculates the interest due and raises a fixed-term deposit note for each customer's deposit. The deposit note is a two-part set: one part is sent to the bank's archives, the other is filed.

Next the order clerk raises a telex to move funds by interbank transfer. The telex is passed to the cables department and then the deposit is recorded in two ledgers: the first is a ledger showing each deposit by customer and by bank, while in a second, deposits are added to each customer's balance (if present) with the receiving bank.

At the beginning of each day old deposits mature, in other words the money which has been on deposit for the agreed amount of time is returned to the FNBR with interest. Matured deposits have to be removed from the ledgers at the start of each day's trading and the customer balances adjusted correspondingly.

Sometimes deposits do not run to their maturity date but are sold to other banks as *certificates of deposit*. In this case the FNBR places money with the other bank as a certificate of deposit which may be redeemed on any day according to the FNBR's wish. This enables money to be redeployed if more favourable market opportunities appear. Apart from the selling details, certificates of deposit are treated exactly as normal deposits. The decision whether to place money as a fixed-term deposit or a certificate of deposit is the responsibility of the head dealer; in practice, fixed-term deposits account for 90–95 per cent of the deals.

When deposits are either sold or redeemed the money is automatically returned to the FNBR. Checking the correct amount is returned on time is the responsibility of the treasury department and is not within the scope of our study. In trading operations, the department we are investigating, the interest is calculated for matured deposits and entries are removed from deposit and balance ledgers before the beginning of each day's trading.

The dealing subsystem

In another interview one of the dealers describes his part of the system.

New amounts of money arrive from customers, via the treasury department, at various times during the day. For convenience the monies are batched into morning and afternoon sessions. Money has to be deposited for a fixed duration (30, 60, 90 days). The funds received from customers are sorted in separate duration groups and placed on file. The groups go into morning and afternoon batches that form a potential deposit to be placed with another bank.

Customers maintain blacklists of banks which impose either a credit limit which their balance with a particular bank must not exceed, or a complete ban on placing their money with a bank. Dealers forecast the effects of placing potential deposits by adding the new monies to customers' balances already held with other banks and then looking to see if any of the customers' credit limits have been exceeded. If so the potential deposit is rejected. Otherwise the dealer adds the bank in question to a list of potential acceptable destination banks. When deposits for all the customers have been forecast the dealer looks for banks which are common to all the individual customer lists. He then phones around these commonly acceptable banks to find out what interest rates are on offer for the amount of money available and then consults his own list of credit ratings of other banks. Generally the bank which offers the highest interest rate with the lowest credit risk is selected – the decision-making is too complex to explain to a mere systems analyst, and is part of the dealer's skill.

Once he has made the decision the dealer confirms the deal by phone to the other bank and fills out a ticket, which he sends to the clerk to record the deposit.

A.2 INITIAL ANALYSIS AND MODELLING (MODELLING STAGE)

On initial examination of the system some entities suggest themselves for consideration. Obviously Dealer and Clerk are two objects in the real world system, which has a problem based upon money. Our first reaction may be to model some of the functions which appear in the narrative, such as the forecast process; we shall resist temptation, however, and look behind the surface functions to find out what the system problem is. A rough guess is that it is about money – finding out where to deposit it, doing so and then retrieving it. Involved with this process there appear to be at least two objects, Dealer and Clerk, and maybe the Customer as well. The system has to respond to the events of receiving money from customers. It next has to react to decisions taken by the dealer and clerk and then to interact with other objects outside its boundary, namely other banks and the treasury department.

Examining the provisional list of entities we suspect that Customer may give rise to a marsupial, as one customer may place several amounts of money with the bank over his or her lifetime. The money will have a separate life history of its own. Also customers appear to do very little, apart from placing money with the bank, and the only actions which can be ascribed to Customer are Place and Receive. In view of this we may choose to relegate the customer to a point outside the system, or machine boundary in JSD terminology.

A good starting point with JSD is to ask what the basic problem is and what are the real world objects to which the system has to react. In this case the problem is about money. To start, we settle for three entities based on the problem itself and two principal actors involved with the problem (Money, Dealer and Clerk) and start modelling. We know that money comes into the system from customers, various things happen to it until a suitable destination is found and then it is dispatched to another bank to stay on deposit for a set number of days. We can be fairly sure that money comes in as a distinct entity (i.e. in discrete amounts) but then it will react to events from the dealer. Its existence becomes more tenuous, because money from different customers gets merged into a larger amount which is a potential deposit.

It appears, therefore, that the real entity we need is Deposit which starts life as a grouping of customers' funds and then goes through a forecast process whereupon, if it survives, it becomes a real deposit which is placed with another bank until it matures and is finally redeemed. Could we use Deposit to model the whole life history from the instant when money arrives from customers? This would be difficult, because Deposit only comes into existence when individual customers' money is pooled in 30, 60 and 90 day groups. This represents a timing clash between actions on one object, an individual customer's money, and another object, the Deposit, which is a grouping of several customers' money; hence we need a separate entity to model the first stage of the system.

Accordingly we downgrade money's role and revert to modelling the customer's actions of placing and receiving funds. The customer life history is restricted to placing money with the system until it is returned again. Now we can turn our attention to the second principal entity. The Dealer appears to go through the steps of finding out where the money should go and then placing it on deposit with another bank. Our dealer entity will have to react to these decision-type events. We can start the system model with Customer and Dealer. We have introduced another entity, Deposit, to model the lifetime of the money once it has been pooled.

We may want to model the forecasting process within Dealer, but we should beware: forecasting looks like a series of actions which also affect the deposit. It may therefore be an interactive function, so we shall leave resolution of this point until we have more detail about the forecasting process. Dealer needs to be connected to Deposit by a datastream which contains the potential deposit for evaluation. Dealer outputs a ticket datastream which will go to the recording part of the system.

The next part of the system appears to be a simple recording process. Inspecting the narrative shows that there appear to be four things (at least) which could be modelled: Clerk, Ticket, Deposit and Balance. We already have a Deposit process so we can add to its life history. Another candidate is the Clerk who, we are told, checks the tickets and transfers the money to the receiving bank. The Clerk's role in retrieving the money is less clear.

We could try modelling the ticket but most actions are done to it by the Clerk; thus, preferring to model active entities, we relegate the ticket to static data. The Clerk entity seems to have good coherent time ordering in the recording part of the system; but, it is not clear whether the redemption of deposits is also part of the clerk's responsibility. Most of the clerk's activities are short-term procedures which are better modelled as interactive functions, and in the light of this we shall exclude the clerk from our initial analysis. We already have an entity modelling the deposit which is affected by the clerk's activity, so we can use it instead. We must remember that we are not modelling people's jobs as we find them; rather we must look for logical groupings of time-ordered actions which respond to an object in the real world system. These may be carried out by one or more people.

Balance entity is more difficult. Maybe we could include it in the Deposit entity? But this would be unworkable because we would discover a marsupial clash: there may be several Deposits contributing to a single customer Balance. Therefore we choose to model a separate Balance entity.

There are some other potential entity candidates such as the treasury department and other banks, but we are given very little information about them except that they are recipients or providers of event messages. As the business of sending and retrieving money between the treasury department and other banks is an interaction external to the system we will not model these objects. Instead we choose to use the entities Dealer and Clerk to respond to them. Consequently we place them outside the system.

Our first attempt has given us four entities: Customer, Deposit, Dealer and Balance.

Phrase listing

To cross-check our analysis so far we can try the alternative approach of phrase listing to see if it suggests the same entities. The following statements are concise versions distilled from interview notes.

Customers place funds with the FNBR
Customers impose credit limit constraints
Funds are grouped by duration
New funds are added to existing balances
Balances are compared with credit limits
Acceptable banks are placed on a forecast list
The list is searched for common banks
Common banks are communicated to dealers
Dealers assess credit risks
Dealers phone for interest rates
Dealers select the most favourable bank
Dealers place money on deposit with banks
Deposits mature after a set number of days
Some special deposits may be sold before their normal duration is complete
Funds are returned to the FNBR
The redeemed amount is deducted from customers' balances
Funds are returned to the customer
Customers place deposits with the FNBR
Customer deposits are grouped into potential deposits
Potential deposits are forecast by dealers
Forecasting compares customer balances with credit limits
Potential deposits are evaluated to select a suitable bank
Dealers evaluate banks by balancing interest rates against credit risks
Deposits are placed with suitable banks
Deposits may be placed as either fixed-term deposits or certificates of deposit
Dealers record deposits on tickets
Clerks partition tickets into customers' deposits
Customers' deposits are added to their balances
Mature deposits are returned to the FNBR
Mature deposits are subtracted from balances.

Analysis of the list gives us a similar list of entities but with some differences:

Deposit	Funds
Dealer	Balance
Customer	Bank (other)
Clerk	FNBR

The main difference is the identification of two bank entities, the FNBR and the other banks; however, the FNBR is made up of entities we already have such as Dealer and Deposit. We may want to model the other banks but this entity is essentially passive, just accepting and returning the deposit. We have already decided that other banks and the treasury are a separate subsystem which we do not wish to model. Therefore we can exclude other banks as outside the system boundary.

One confusing feature of the phrase listing is the multiple use of the word deposit. As long as we have a clear view that 'deposit' can be two separate things our modelling will be correct. The two objects are the customers' money which arrives in the system, and the grouping together of this money forming a deposit to go to another bank. Clerk may be an entity but we have already rejected it as a function; customers, funds and other banks may be possible candidates, however. Forecasting could also be mistaken as an entity in its own right unless the analyst was wary of functions. Funds and other banks are passive objects which do not create any events of interest. We therefore reject them, leaving four entities: Customer, Dealer, Deposit and Balance.

Action attributes

Before drawing up entity action lists we can cross-check our results so far by looking for action attributes, i.e. the external events to which the system is going to have to respond. Customers input money into the system, so arrival of money forms one event. Dealers transmit several decisions as events which will affect the fate of the deposit, e.g. rejecting or accepting forecasts, placing deposits, creating and amending tickets. The Deposit and Balance entities will respond to the events of being forecast, accepted, placed, redeemed and so forth.

After a preliminary analysis of the external events which may give rise to action attributes, the next step is to draw up the entity action list.

Entity action list

Entity: Dealer

Actions	Attributes
Gather-info	Bank-Ratings, Forecast, Interest rates
Consult	Strategy = (Cert Dep or Fixed term)
Forecast	Bank balances, Limits
Evaluate	Forecast list, Ratings, Interest rates

Place	Deal, Value, other bank ID
Record	Deal, Date, other bank
Amend	Ticket details
Confirm	Deal, Value, other bank ID, Telex
Abandon	Deal, Forecast list

ACTION LIST

Gather-info. Dealer assimilates information necessary to evaluate the deal – bank credit ratings, the forecast of acceptable banks and interest rates on offer.

Consult. The head dealer is consulted to decide if a fixed term or certificate of deposit should be purchased.

Forecast. Customers' potential balances are compared with credit limits.

Evaluate. The deal details are evaluated, weighing up the interest rates offer versus the risks.

Place. If the dealer decides the deal is favourable he places the deal with the other (destination) bank over the telephone.

Record. Confirmed deals are recorded on a ticket which is sent to the order clerk.

Amend. Errors on tickets are amended at the request of the order clerk.

Confirm. Once the deal has been placed it is confirmed by telex.

Abandon. Unfavourable deals are rejected and a new forecast list is ordered.

The dealer entity models the decision history which results in customers' money being placed as deposits with other banks. Many dealer actions may not be suitable for automation, as the narrative emphasizes that choosing which deal to do is a matter of human judgment and experience. As such there is a case for considering Dealer to be outside the machine boundary; but our user has stated that the area of the dealer's activity should be analysed, and we therefore include it in the list.

Entity: Deposit

Actions	Attributes
Grouped	Start date, Value, Customer ID
Check	Customer ID, Value
Accepted	Dealer decision
Failed	Dealer decision
Placed	Start date, Bank ID, Amount
Matures	Maturity date
Removed	Deposit note ID
Sold	Certificate of deposit ID, Sold date
Copied	Deposit note details
Archived	Deposit note details

ACTION LIST

Grouped. Monies from several customers are grouped to form a potential deposit.

Check. The potential deposit is checked to make sure it does not infringe any clients' restrictions.

Accepted. The potential deposit is cleared to be placed.

Failed. The potential deposit is rejected because it does infringe clients' restrictions.

Placed. Deposit is entered in the FNBR's records.

Matures. When the maturity date is reached the deposit is redeemed from the other bank.

Removed. Matured or sold certificates of deposit are removed from the FNBR's records.

Sold. On instructions from the head dealer certificates of deposit are sold and the money redeemed.

Copied. All deposits which are removed are copied to the financial division to update the general ledger system.

Archived. All removed deposits are also copied for the FNBR's archive.

Deposit records the fate of the main object in the system. It starts life as a grouping of customers' money which forms a potential deposit. This is then submitted to forecasting which decides if it infringes any customer credit limits or not. If it does then the potential deposit is rejected; otherwise it is placed as a fixed-term deposit or certificate of deposit with another bank. Two entity processes could have been modelled, as Fixed-term deposit and Certificate of deposit, but the narrative stresses that they are treated exactly the same apart from one being sold; therefore merging the two is appropriate.

Entity: Balance

Actions	Attributes
Opened	Customer ID, Bank ID, Amount
Closed	Customer ID, Bank ID, Amount
Increased	Customer ID, Bank ID, New amount
Decreased	Customer ID, Bank ID, Subtracted amount

ACTION LIST

Opened. If no money is held by a customer with the destination bank a new balance is opened.

Closed. If a balance is reduced to zero it is removed.

Increased. New deposits are added to existing balances.

Decreased. The value of sold certificates of deposits or matured deposits is subtracted from existing balances.

Balances are held with other banks by the FNBR on behalf of its customers. Balance describes the life history of the aggregate of a customer's money placed with another bank. An alternative model could be either Customer or FNBR, but as many balances are held with other banks and each customer may have several balances, the balance life history is likely to be independent of customer and FNBR actions. We would have to separate Balance from Customer as a marsupial process. In this case we have recognized the marsupial sooner rather than later.

Balances will undergo process creation and deletion as new customers are added and money is placed with new banks. Potentially there could be an instance of this entity type for every customer–bank combination; in practice, there will be fewer instances because some customers operate credit bars on certain banks.

Entity: Customer

Actions	Attributes
Place	Money amount, Currency, Customer ID, Duration
Receive	Money amount, Currency, Deposit ID

ACTION LIST

Place. Money is placed with the FNBR for a set number of days.

Receive. The money is received back after being on deposit.

Customer is a fairly trivial entity which describes the initial fate of the customer's funds when they are handed over to the bank. There may be more complications that we are unaware of (such as the funds being amended or cancelled) but so far we have no knowledge of this.

Other potential entities

Other potential candidates for entity processes are Bank-Ratings, Customer-Limits and Forecast. Limits and Ratings look like attributes of Customer and Other banks; but we have very little information about these processes and their time ordering, and therefore Rating and Limits are regarded as outside the system boundary. Limits, however, may be changed by customers so we may need to include them later in analysis. Forecast is currently described in Deposit but we have not yet analysed details of the forecasting process. If the Forecast actions are complicated and form a separate time ordering apart from the deposit it may have to be separated off as an interactive function.

A.3 ENTITY STRUCTURE ANALYSIS (MODELLING STAGE) AND INITIAL MODEL (NETWORK STAGE)

Using the entity action list we can start modelling entity structures to produce model processes. Within the system we have four model processes, Dealer, Customer, Deposit and Balance; as well as these there are several processes outside the system boundary: Other-Bank, Treasury and Credit-Ratings. Processes external to the system may either act as sources or destinations of data, or have information in their state vectors which model processes need to access.

Model process: Dealer

Dealer is an iteration of actions which are organized in two subtrees: one does the dealing, the other process queries tickets. These are selections, as we have no way of knowing which the dealer will do next – this depends whether he has a Potential-deposit or a Query in his in-tray.

The deposit trading subtree is a sequence of gathering information, forecasting the deposit, evaluating the deal and then placing the deposit. The Evaluate action depends on a datastream message bearing the reply from a real world dealer. If the deal is acceptable then Record is the action in this sequence which writes a ticket datastream. The other branch in the selection requests errors in tickets to be dealt with (see Figure A.1).

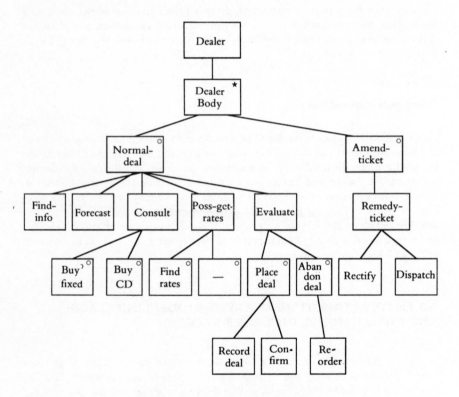

Figure A.1 Model process: Dealer.

Model process: Customer

Customer (Figure A.2) is a fairly trivial process which places new funds with the bank and then receives funds back when the deposits have matured.

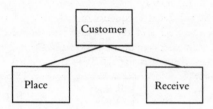

Figure A.2 Model process: Customer.

Model process: Deposit

Deposit (Figure A.3) has input datastreams from Customer and Dealer. These datastreams will tell Deposit when it has been placed and removed. Deposit is a sequence of being grouped, then forecast (which it may fail), being placed if acceptable and finally being redeemed either as a sold certificate of deposit or as a mature fixed-term deposit.

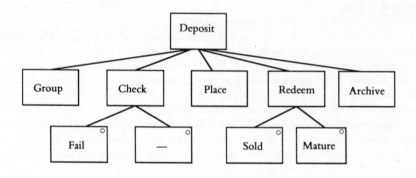

Figure A.3 Model process: Deposit.

Model process: Balance

Balance (Figure A.4) has an input datastream from Deposit which either adds or subtracts money from a balance value. The process starts with the balance being opened; this is followed by a lifetime of zero or more changes and ends with the balance for a customer with a particular bank being closed if no money is left on deposit.

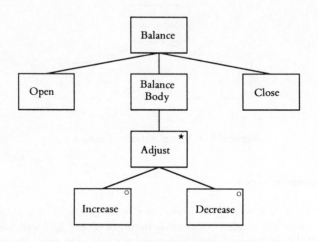

Figure A.4 Model process: Balance.

Initial model

Dealer has to respond to events caused by the deposit recording part of the system ('qry' messages for errors on the tickets) and events from the real world dealer which decide the fate of a potential deposit. Besides these inputs the Dealer also receives a forecast (fcst). Customer has one input, new funds being placed with the bank (fnd), and two outputs, one passing the funds on to Deposit (pdp) and the other returning the money to the external world customer (rdp).

Deposit receives input from Customer as a potential deposit (pdp) and then gets instructions from Dealer which governs its fate as a placed deposit etc. (pld). It outputs one datastream (ajbal) to Balance, to which it is the only input. The initial model is illustrated in Figure A.5.

A.4 A REVIEW OF PROGRESS AND FURTHER ANALYSIS

After we have started the initial model it becomes obvious that we are missing details of the dealer's operation. Dealers are busy people and rarely give their time generously, so we examine the PSDs and SSD we have already constructed to prepare a list of questions. In the Dealer process, we do not have an accurate picture of the forecasting process. How many limits can be exceeded, if any, before a potential deposit is rejected? If a potential deposit fails the forecast we need to know its subsequent life history. This raises further questions of sequence control; if a potential deposit fails, is it recycled immediately or does it go into a queue with other incoming deposits?

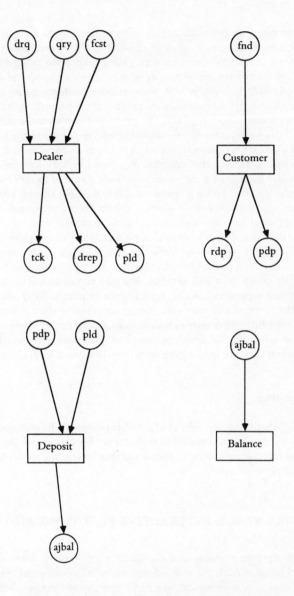

Figure A.5 Initial model: Deposit dealing system.

Another question concerns the sequence of actions in placing deals and correcting errors in tickets. We have assumed this is done during the day, but the dealer may have a specific time for correcting errors – maybe at the end of each day.

Forecasting system

A further interview with the dealer reveals what happens in the forecasting process.

Customers may exercise two types of constraint, sometimes laying a complete ban on a particular bank or setting a monetary value (expressed in US dollars) which their balance should not exceed. The value of the customer's money placed with the FNBR is added to any balances held by that customer with other banks. The banks which are permissible for the customer are noted and the process repeated for all the customers in a potential deposit group. The permissible banks for each customer are compared and the banks common to all customers are put on to the forecast list.

Sometimes there are no banks that are acceptable to all customers, in which case the batch is rejected. The rejected potential deposit is then split into separate groups by dividing 'easy' from 'awkward' customers using a rule of thumb based on their limits. The easy customer group is reprocessed as before, the rule of thumb ensuring that this group never fails twice. Awkward customers are processed individually, which ensures that at least one bank is acceptable to them.

We also discover that dealers may reject a potential deposit even after it has passed the forecast test. If only one or two banks are acceptable they either may offer poor interest rates or may be a bad credit risk in the dealer's experience. The dealer can ask for the customer group to be split as before into easy and awkward customers which are then processed separately. Dealers tend to process potential deposit groups that have passed the forecast and easy customer groups before they process awkward customers. Correcting ticket queries is their lowest priority and is often carried out late in the day when all the deals are complete. Forecasting is ususally done by dealers, but when they are busy clerks may prepare deposit forecasts for them.

Deposit recording

The knowledge that limits are held as US dollars prompts a further question about how different currencies are handled in the system. We check with the clerk, who tells us that the forecasting is done in dollars and that currency is converted to dollars before it is added to balances.

A.5 NETWORK STAGE: INTERACTIVE FUNCTIONS AND THE SSD

Having drawn up entity action lists and made a preliminary draft of the entity structures and initial model, we realized that more information was required. We now have the necessary information, so entity structures and the initial model can be refined in the light of facts gained from our further analysis. We are now in a position to add interactive functions and make a first draft of the system network. We shall start by separating out the forecasting process as an interactive function which communicates with Dealer by two datastreams, fcst and drep. We shall add the clerk's job of checking the tickets as an interactive function. Also, input from outside the system will be added to the Dealer and Clerk processes. This input will be messages from the people operating the system, i.e. the dealers and clerks. First, the Dealer model process must be modified as a result of these changes.

Dealer

We did not have enough detail of the dealer's actions in the initial narrative to describe the dealer's time ordering in detail. From the next set of analysis notes, however, we learn that the dealer goes through a sequence of receiving a forecast list, deciding whether to buy fixed-term deposits or certificates of deposit, possibly getting interest rates offered by other banks and then deciding whether to place a deal or not. It is apparent that dealers attend to normal deals first, followed by deals for awkward customers and finally process queries from the order clerk when they have spare time.

Dealer has three major possible action branches: either doing a normal deal or doing an awkward customer deal or amending a ticket. The input datastreams (fcst, qry) are rough-merged. The fcst datastream is the one we expect to process most frequently, with queries arriving less often. For the time being we shall not differentiate between initial forecasts and later ones for awkward customers; all will be input on the fcst datastream. Normal-deal and Awkward customer deal share a common subtree which is a sequence starting with Find-info and finishing with an Evaluate action. The dealing strategy is selected by consulting the head dealer about whether to buy a fixed-term deposit or certificates of deposit. This is followed by a Poss-get-rates action which models the dealer's probable intention (communicated on the datastream drq) to phone for new interest rates. The Dealer then Evaluates the deal and depending on input from the real world dealer (drq) the deal is either accepted or rejected, in which case a new forecast is requested.

The final part of the selection, Amend-ticket, deals with queries which may have accumulated in the input datastream. Details to answer queries are read from qdt and amended tickets are written to tck (see Figure A.6 for the revised Dealer PSD).

Order-clerk

Order-clerk has two main subtrees: one removes the mature and sold deposits by inspecting Deposit state vectors, the other deals with checking the tickets produced by the dealer. We know that deposit removal is performed at the start of each day, so a sequence is modelled of removal followed by validation of many tickets.

Tickets may either be correct or fail the checks, depending on qualities of the data which can only be determined at run time. Therefore instead of modelling a process structure with a series of nested selections, a backtracking structure is used. Backtracking has been used positing a 'good ticket', quitting if it fails the checks and admitting a bad ticket. The ticket can fail at either the Check IDs or the Sense-check step (depending on the clerk's reply crep); in both cases the sequence is halted and side effect action is taken to write an error message to the dealer (qry) after getting the explanatory comment from the external world clerk via crep.

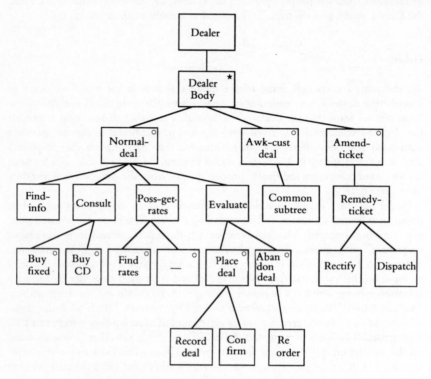

Figure A.6 Model process: Dealer – amended version.

If the ticket passes the checks the remaining components, Partition and Record-deposit, are fairly straightforward. Before adding the new deposit to the balance the currency is Converted to US dollars (Figure A.7).

Another review and more problems

On further inspection the forecasting system still presents problems. We know that if a single customer fails the forecast the whole batch has to be rejected and reforecast after dividing the customers into different groups. Also the whole batch may be rejected by the dealer. A potential deposit therefore may become two or more real deposits if it is rejected. This creates a timing clash, because we cannot model what is happening to the separate life histories of the awkward and easy group deposits in the same diagram. The potential deposit has spawned a marsupial entity, the real (i.e. placed) deposit which may be derived from either an easy or an awkward customer group.

If we did not split the potential deposit process to cover the contingency of a rejected forecast, further processing of potential deposits would present timing

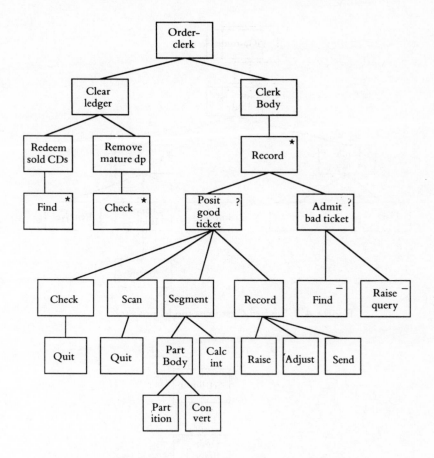

Figure A.7 Interactive function: Order-clerk.

difficulties. We have discovered a timing clash between the demands of the original Deposit and the reforecast system; accordingly the Deposit is split into two separate parts, Forecast-deposit and Deposit. Forecast-deposit becomes the actions carried out on the potential deposit to decide whether it passes the test or not. This looks like another interactive function.

At this stage is seems appropriate to revise our judgment about the Customer model process and add actions for the setting of limits with banks, either complete bans or cash limits. The revised model is depicted in Figure A.8. Unfortunately the model is incorrect because customers have limits with many different banks; the subtree of actions for setting limits is therefore a repeating group within Customer. We have discovered another marsupial. The Customer process is dismembered into the original model process and the Set-limits process which models the lifetime of the customer's constraints (Figure A.9).

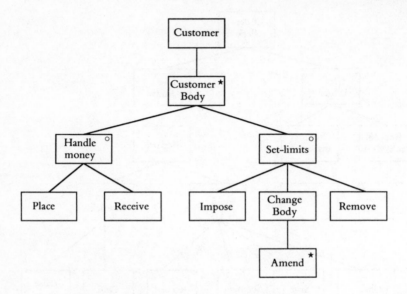

Figure A.8 Revised Customer model process.

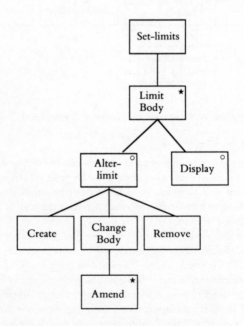

Figure A.9 Interactive function: Set-limits.

Deposit

The revised Deposit model process has a sequential life history of being Placed, Recorded, Matured, Removed, Copied and Archived (Figure A.10). If we choose to model Fixed-term deposits and Certificates of deposit as one process, then another component has to be added to account for the possibility that a certificate of deposit may be sold. We now need to clarify the role of this entity. Deposit can be viewed as one single large sum of money placed with another bank, and this is indeed the recipient bank's view. But within the FNBR a deposit may be viewed either as one sum or as a group of several deposits made by individual customers. We shall choose the latter view, because this will provide more detailed information about the grouping of money which is necessary for the forecasts.

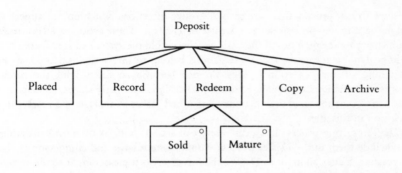

Figure A.10 Model process: Deposit – amended version.

The Deposit model process has no way of knowing which event will happen next; consequently a rough merge of input datastreams (pdp, rdp and pld) is needed because the order of input cannot be predicted. Resolution of the ordering will have to wait until implementation, in which case the Scheduler will access the Deposit process state vector to ensure the correct action is taken. The alternative is to add a further process to order the rough merge messages, but at this stage we can tolerate the lack of control over input timing.

Balance

Balance appears to be a sequence which is opened, may be changed several times as new deposits are added and old ones deleted, and will eventually be closed. Balance receives datastreams from Clerk in a similar unpredictable manner as Deposit does, and each message could require either an add or subtract action to take place on a balance. As the Balance model process cannot be sure which action will happen next, we could model it with a selection depending on the content of the datastream ajbal. Depending on which record is present the balance will either increase or decrease. Opening and closing of Balances will have to wait until implementation. State vector inspection will be required to determine if the input message represents a new balance or refers to an old one. This is an example of process creation and deletion.

The uncertainty due to message handling could be tackled by imposing a three-way selection (Open, Update, Close) on the model process structure, but this destroys the time ordering in the model and impoverishes its information content. A better policy would be to keep the original time ordering of Balance, which reflects its life history, and leave the message handling to implementation. At that stage the Scheduler will accept the messages, access the Balance state vector and then by means of the text pointer run the appropriate bit of balance code, to open, update or close a balance.

Interactive function: Forecast

In the Forecast process there are several groups of actions which do not appear to belong in a model process such as Dealer or Deposit. These actions are concerned with changing the qualities of the basic entities within the system such as Deposit. It is important to recognize that these actions belong to interactive functions at the beginning of analysis so that they do not become confused with the model processes. Functions are not added to the JSD design in the early stages, so suspected interactive functions and their actions were noted earlier in analysis and marked for inclusion in this step.

The forecasting process has a discrete set of actions with its own time ordering. To include them in the Dealer process would make it large and cumbersome; also forecasting creates an input to Dealer, the forecast list. It creates input to the system which does not come from outside the system; hence it appears that forecasting is an interactive function. Consequently we do not wish to elaborate the forecasting process within Dealer, and instead we create the interactive function Forecast which interacts with Dealer. The process structure of Forecast is elaborated in the following sections.

System network: first draft

The first draft of the system network can now be prepared, and is shown in Figure A.11. This will contain only the core of the system (model processes and interactive functions) and is intended to clarify our analysis.

Customer feeds Deposit with new funds from real world customers. The interactive function Forecast inspects state vectors of the Set-limits, Balance and Customer processes to determine the forecast of potential deposits acceptable to customers. It passes this forecast to Dealer in the fcst datastream. Dealer takes the forecast and responds to input from the external world dealer (drq) and then either rejects the forecast by sending a negative drep message to Forecast or sends a positive message to Forecast and a pld message to Deposit.

Dealer also sends a tck message to the Clerk function which validates it and sends any errors back on the qry datastream. The Clerk process then writes a datastream to update the balance. The Clerk also removes balances and matured deposits by writing rdp and ajbal messages.

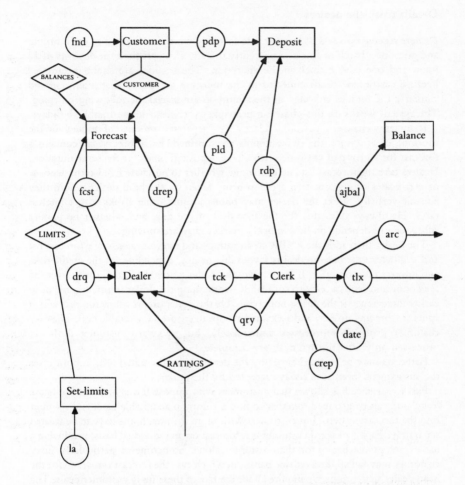

Figure A.11 Deposit dealing system: system structure diagram (first draft).

A.6 USER REQUIREMENTS FOR THE PROPOSED SYSTEM

So far we have gained a reasonably accurate picture of the current system which we have modelled, but our user has not given us any requirements as to how the new system should work. Also we still need more information about time ordering in Dealer and Order-clerk. For instance, how do the dealers select awkward customers? And if customers' money is forecast in a batch-type operation, what do dealers do on Monday mornings while they wait for the arrival of customers' funds? The following narrative is synthesized from interviews and meeting notes after further visits to the dealers and clerks.

Details from the dealers

Dealers receive two sets of forecast lists each day which correspond to the morning and afternoon batch of funds. Apparently there is a cut-off time for deposits of 12 noon and one of 4 o'clock in the afternoon. The sessions are slight misnomers because customers' funds collected in the morning are placed that afternoon and similarly the monies collected in the afternoon are placed the following morning. The system works on this phase lag principle over weekends and public holidays.

If by any chance a customer places two separate amounts of money in the morning or afternoon the two amounts are summed. Each forecast list contains a forecast for a group of customers who have deposited funds for the same duration. Dealers take the forecast list and determine whether to purchase fixed-term deposits or certificates of deposit, this decision being taken by the head dealer whom they consult verbally. Next the dealer may phone around other banks to get interest rates. He always does this for the first deal of the day, and whether he phones subsequently depends on how quickly market rates are moving.

The dealer then adds the FNBR credit ratings and interest rates to the forecast list and evaluates whether the deal is favourable or not, depending on the credit risks and interest rates on offer. If favourable, the dealer places the deal verbally over the telephone and sends a telex to confirm it – something the dealer omitted to tell us in earlier interviews. If the deal is not favourable the dealer orders a new forecast which splits customers into easy and awkward groups to ensure a favourable deal. The easy customer group is reprocessed immediately, but awkward customer deals are processed only after normal deals are complete.

Forecasts may be rejected either by the dealer or by the initial test; in both cases the subsequent forecast is always accepted by the dealer.

Past experience has shown that customers who impose fewer than ten limits are 'easy' and can be grouped together to find a common acceptable bank and re-input into the forecast system, but customers with more than ten limits have to be treated separately. Each awkward customer is reforecast on an individual basis, producing a list of acceptable banks for that customer alone. Sometimes a particularly fussy customer may fail all the available banks, in which case the forecast is marked for the head dealer's attention. We inquire about the fate of these fussy customer deals. The dealer informs us that the outcome is usually for the head dealer to phone the customer and point out that his limits are excessive. This involves unpredictable negotiations with which we need not concern ourselves.

The whole dealing process has to be co-ordinated so that one deposit lifetime group is completed before another group is started, although single awkward customers are allowed to accumulate until the end of the deposit sessions.

Some tickets are returned by the order clerk with queries. These are dealt with when the dealer is not busy with other matters. The erroneous fields are corrected and the ticket initialled as being correct before being sent back to the order clerk.

Further details from the order clerk

The time ordering in the Clerk model process is reviewed with the order clerk, who makes the following comments. Each order clerk services several dealers, and tickets therefore tend to pile up in his in-tray. The clerk takes each ticket, checks it against reference lists of customers and banks and then sense-checks the rest of the data. Any queries are marked in red and an explanatory note is added on the bottom of the ticket which is returned to the dealer. If the ticket passes the checks, the clerk processes it normally.

Deposits which have matured and sold certificates of deposit are removed at the start of each day's trading. Balances are adjusted at the same time. New deposits are added to balances throughout the day; the time lag between placing a deal and recording the addition to the balance is accepted as tolerable as long as all deposits are placed and recorded during the same day.

User requirements

After a further analysis session with the head dealer we establish which functions our user wishes to automate, and how. Although some of the dealer's actions appear to be nonautomatable, our user wants a dealer support system which will provide information for the dealers to help them make decisions and which will communicate between dealers, the head dealer and the order clerk using electronic messaging to transfer lists, tickets and so on. Interest rates, banks and customer IDs, together with the head dealer's strategy decision, are all held on another system but may be accessed as external state vectors. The list of requirements is as follows.

Dealers

- To be able to receive the forecast list automatically and display the list, selecting either normal or awkward customer lists.
- To add FNBR bank credit ratings to the list.
- Input, store and display interest rates being offered by other banks.
- To sign off a deal and automatically generate either a ticket and confirmation telex or a reorder forecast instruction.
- Edit queried tickets.
- Because of the unpredictability of a busy dealer's day each function should be available on demand from a menu.

Order clerk

- Automatically display tickets, mark errored fields and add comments explaining errors.
- Generation of deposit notes, archive copies and telexes.
- Automatic updating of balances.

Customer funds and deposits

- Automatic elimination of mature deposits at the start of each day and adjustment of the appropriate balance.
- Same process as above for any sold certificates of deposit.
- Automatic sorting into lifetime groups, display of money in groups and capacity to sort into awkward and easy groups before invoking the forecast system.

Customer limits

■ Ability to add, delete, display and change customer credit limits on demand throughout the day.

Ratings

■ Ability to add, delete, display and change FNBR bank credit ratings on demand throughout the day.

Balances

■ Ability to view customer balances on demand, accessed by customer ID.

Deposits

■ Display deposits by duration, customer ID or reference number.

A.7 ELABORATION OF THE SYSTEM NETWORK (NETWORK STAGE)

From the user requirements and further information gathered during analysis we are now in a position to elaborate the system further. We shall investigate three areas of the system in turn, reviewing first the system model and interactive functions, then the information subsystem and finally the input subsystem.

Interactive functions

From further analysis it appears that the Forecast process is considerably more complex than we first thought. When forecasts are rejected we need to divide the customers up into easy and awkward groups, something which we so far have not modelled. We could try modelling the life history of a successful and an unsuccessful forecast, but then we have the problem of when to decide that a forecast is unsuccessful. We need something to model the time ordering of an awkward forecast and a process to co-ordinate the activity of the subsystem. These processes are all short-term procedures which will create feedback input for the system, so they are also suitable for modelling as interactive functions. The processes each have a discrete set of actions with their own time ordering, provide input to the Dealer model process, and carry out a task within the system by interacting with Dealer.

A new process Divide-customers is added which splits a failed forecast group into easy and awkward groups for input into the Forecast function. Easy customers are treated exactly like any potential deposit and can be re-input into the Forecast function. Awkward customers are treated differently so this time ordering will be modelled in the Divide-customers function.

Control of the forecasting system has now become quite complex. We could introduce a Clock process to co-ordinate Forecast, Money and Divide-customers, but there is little information about the exact timing of these processes. All we know is the sequential dependency that if the forecast fails then easy customers are reforecast before the awkward ones. Therefore we shall introduce another inter-active function Select-group to co-ordinate activity of these processes. Select-group will be responsible for picking potential deposits one at a time and feeding them into the Forecast process. If the forecast fails it will invoke Divide-customers to split the groups and reforecast them before continuing with the next potential deposit.

In the recording part of the system, part of the Clerk's job is to remove matured deposits. This is a series of actions which form a discrete sequence of their own and therefore forms another good candidate for an interactive function. We could keep the actions within Clerk, but the run time execution of the two parts of the Clerk process appear to be quite different. In the interests of optimization the Clerk process can be dismembered into two separate interactive functions. We know that the Remove function is invoked at the start of the day but it may not always be so and this task may be carried out by someone else besides the Clerk. This gives a separate time ordering, which reinforces the decision to make it an independent interactive function.

In the user requirements we have been asked for the ability to change customers' limits and bank credit ratings. This will entail modelling one new process, Rating, as an interactive function. Changing limits has already been modelled in the Set-limits process. The other user requirements for listing of deposits and balances will require imposed functions as the changes necessitate more than addition of a simple elementary operation to model processes. These will be information system functions which do not interact with model processes.

INTERACTIVE FUNCTION: FORECAST

Forecast (Figure A.12) describes how one or more banks are selected as a common destination for a batch of funds. It accepts a potential deposit in one of the three lifetime groups and creates for each customer a series of projected balances by adding new funds to the customer's existing balances. The projected balances are compared with the customer's limits; if they pass then the bank(s) is written to a list of acceptable banks for that customer. When all the customers have been tested, each list is examined to find any banks which appear on all lists. If the 'acceptable bank' list fails to provide a common acceptable bank for all customers the group has to be divided into easy and awkward customers and reforecast.

The input datastreams (esy-cst) and (cstfnd) are rough-merged because Forecast has no way of knowing whether it is doing an initial forecast or a reforecast of customers after eliminating the awkward ones. In practice only one of the input datastreams will have records in it because the Select-group process, which controls the input, waits until a Clear record arrives before filling the cstfnd stream.

Forecast adds the new funds to existing balances, by inspecting the BAL state vector for the customer and comparing forecast balances with the customer's limits (LMT state vector). If the result is positive the money and bank ID are stored in local variables to accumulate a list of potential common acceptable banks. The local variables, a customer x bank matrix, are then compared in the second pass subtree to see which banks are acceptable to all customers, and common acceptable banks are written to the fcst datastream. If no common banks are found a 'none' message is

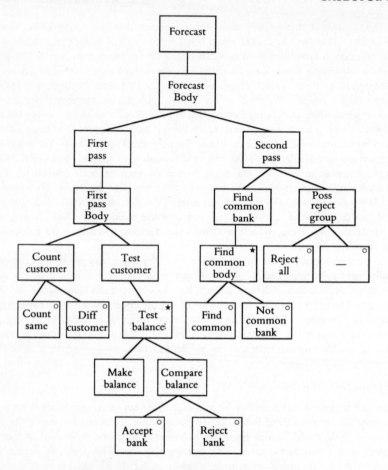

Figure A.12 Interactive function: Forecast.

written to the rslt datastream so that the Divide-customers process can split the customers into easy and awkward categories and the Forecast process can be repeated.

This processing uses internal variables and updates the Forecast state vector with acceptable banks. The Divide-customers process would be difficult to model within Forecast because of the different time ordering when two or more potential deposits may be created from one original. Hence another interactive function is required because data have to be buffered before they are reforecast, thus effectively separating the processes.

INTERACTIVE FUNCTION: DIVIDE-CUSTOMERS

The potential deposit group is split into easy and awkward customers by Divide-customers depending on the number of limits each customer imposes. Divide-

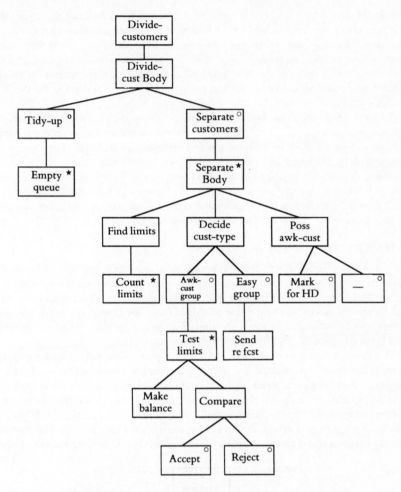

Figure A.13 Interactive function: Divide-customers.

customers (Figure A.13) gets a copy of each potential deposit group. Easy customers will be passed back to the Forecast process, while awkward customers are processed individually. Divide-customers is informed about the fate of potential deposits by both the Forecast, which sends a rslt message, and the Dealer (drep message). If either message is negative the potential deposit has failed and the group will be split, otherwise the copy of the group in the input queue is discarded.

Divide-customers uses a rough merge for reading drep and rslt because it cannot tell whether the dealer or the initial forecast will reject the forecast. If either rslt or drep is negative (= Reject) fndcpy is read and customers divided according to the number of limits they hold and written to two output datastreams, esycst which is fed back into the forecast process and awkcst in which each customer is forecast separately. The awkward customers forecast list (awkcst) is sent to the dealer only after the next rslt message is received from Forecast signifying that the easy group has been passed to the dealer. It is pointed out to us that the dealer may reject an 'easy' forecast group; however, further analysis confirms that dealers always have enough banks to choose from in easy groups and do not reject these forecasts. Banks that fail the 'awkward customer' forecast are written to the rej datastream, and if all banks fail for one customer the output is marked for the head dealer's attention.

To ensure that the input buffer of fndcpy is in step with the messages from the Dealer and Forecast processes, the Divide-customers process clears its input buffer irrespective of whether the forecast was successful or unsuccessful. Therefore if the rslt and drep messages are positive Divide-customers discards its copy of the potential deposit. Only when Divide-customers gets an Accept message in drep and rslt does it know that the group has passed the Forecast process and the Dealer, and only then does it clear its input buffer and write a Clear message to Select-duration to trigger delivery of the next customer group to Forecast.

INTERACTIVE FUNCTION: SELECT-DURATION

This process (Figure A.14) feeds Forecast with potential deposits and controls synchronization of the whole Forecast subsystem. Select-duration waits for a TGM before becoming active; it then runs until its input datastreams are empty but it may halt in several places awaiting Clear messages from the Divide-customers process, which tells Select-duration when the preceding group has completed the Forecast and Dealer sequence.

The Forecast process must consume one group at a time before starting on the next. This could be effected by putting a complex rough merge in Forecast; however, it is simpler to use a separate input control process. Groups are fed into Select-duration in no particular order, so a rough merge of the three input datastreams is used. Select-duration finds which input stream has records present and reads one group at a time, makes a copy, transfers the groups to the Forecast and Divide-customers and then waits for a Clear message before sending the next group.

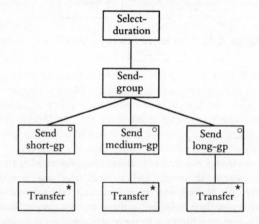

Figure A.14 Interactive function: Select-duration.

INTERACTIVE FUNCTION: START-OF-DAY

To solve the problem of removing all mature deposits at the start of each day an interactive function is added to inspect Deposit's state vectors and to write datastreams to Balance and Deposit to remove matured and sold deposits. Start-of-day reads date and sdep datastreams in a fixed merge with the date first because we want a date message to trigger activity at the start of each day.

Start-of-day (see Figure A.15) searches through the Deposit state vectors, comparing the maturity date with the current date to determine if a deposit should be removed. When a match is found the deposit instance is removed and a message (ajbal) is sent to Balance to subtract the amount from the instance of the Balance state vector representing that customer and bank. This function also searches through and removes certificates of deposit that have been sold.

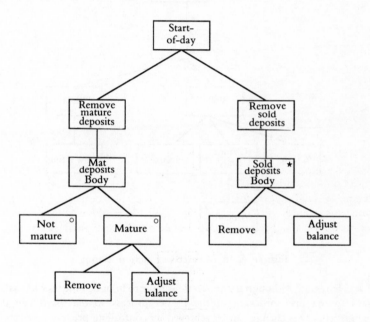

Figure A.15 Interactive function: Start-of-day.

INTERACTIVE FUNCTION: RATING

Changing the bank ratings is another user requirement implemented as an inter-active function, shown in Figure A.16. It is essentially modelling part of the life history of the other banks, in this case only their credit ratings assigned by the FNBR. It is not an interactive function in the strict sense of the definition because Rating does not feed events directly into the system. It does so indirectly, however, by maintaining a state vector file which holds information of interest to the system.

Information system: functions

We need to create displays for deposits and balances and for these requirements new functions, Deposit-list and Balance, will be added to the system. In these cases we shall have to create new function processes because new actions would have to be added to the initial model process – a change which JSD does not allow.

EMBEDDED FUNCTIONS: CHANGES TO OLD PROCESSES

The rest of the user's requirements can be accommodated by changes to the initial model processes Dealer and Deposit, together with the interactive functions Order-

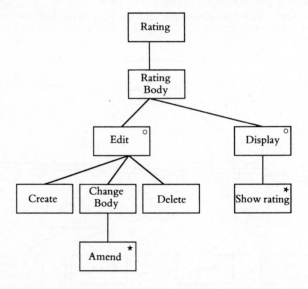

Figure A.16 Interactive function: Rating.

clerk and Forecast. Although no new actions have to be added the user has asked for changes to the time ordering of the Dealer process so that parts can be run independently. This change can be achieved by controlling process execution with text pointers in the implementation stage.

INFORMATION SYSTEM: IMPOSED FUNCTIONS

The user requirements specify two displays, one for balances and the other for deposits.

Imposed function: Balance2
This is a simple process with one action, Display-balance (Figure A.17). A state vector connect is used to link this function to the Balance model process as snapshots of the customers' balances will be required.

Figure A.17 Function process: Balance2.

Imposed function: List-deposit
This function (see Figure A.18) displays deposits in three different ways: all deposits for a certain customer, all deposits of a given duration, or in order of the reference number. This is a three-part selection depending on the type of display requested in the user request datastream (da). Search details are input via depdtl; for deposit-duration and customer searches all the Deposit state vectors will have to be accessed, while for a reference number display it is assumed access will be direct by the key field. Deposits for display are output on deplst.

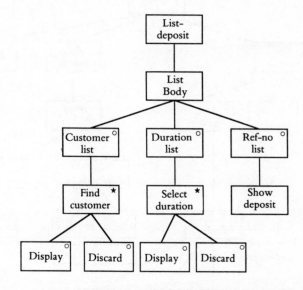

Figure A.18 Function process: List-deposit.

Input subsystem

Filter processes will be required where data are passed across the system boundary. Because the user requires most facilities to be available on demand, separate context filters have not been implemented. The user will supply the text pointer value to run the appropriate part of the system process. The input processes are as follows:

■ A dealer filter accepts replies given by the real world dealer (drq), as decisions whether to accept a deal or not, and other messages answer the clerk's queries about tickets. A Dealer filter (Figure A.19) implements this dialogue and validates the replies. Two levels of menu are implemented. The first requests the dealer to choose between dealing and editing tickets. If dealing is chosen the dealer is requested to select a forecast and then the dealing submenu is presented. This allows the user to run different parts of the process (get information, interest rates, deal and answer queries) on demand.

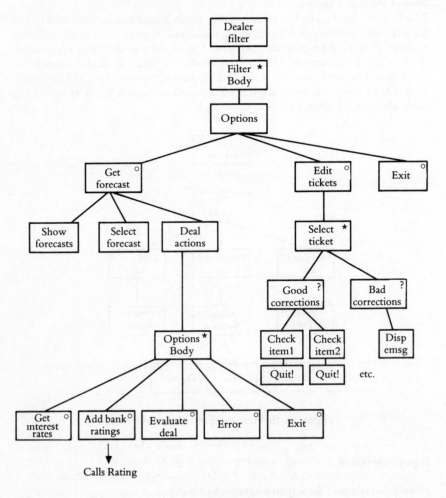

Figure A.19 Dealer filter process.

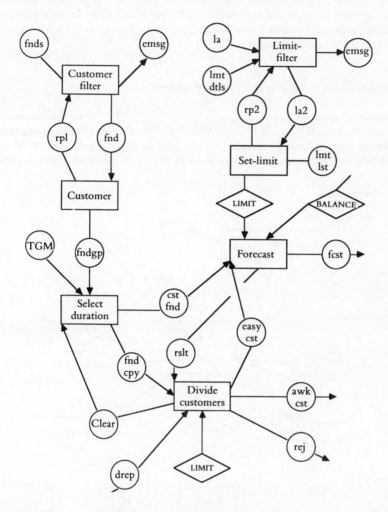

Figure A.20 System specification diagram: Forecast subsystem. System network stage.

- The Query subtree asks the dealer to select a ticket and then inputs correction details.
- The Clerk process accepts input from the real world clerk as decisions about sense-checking the ticket, and possibly raising a query about it.
- Customer accepts customers' funds and Set-list accepts blacklist details from outside the system. These inputs have to be checked.
- The Rating process accepts changes to bank ratings which require validation.

Filter processes will be backtracking structures which Posit normal input, Quit if an error is found and Admit if something invalid has been found. The side effects are intolerable so some action has to be taken, such as issuing an error message and prompting the user to re-input.

Changes to the system specification diagrams

Addition of extra interactive functions with the input and output subsystems has changed the system network since the first draft. The Customer and Deposit model processes are no longer connected so the pdp datastream has been eliminated. Updating the Deposit process is now done by the Clerk and Start-of-day functions

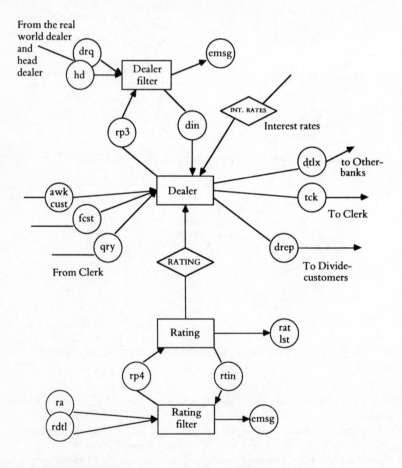

Figure A.21 System specification diagram: Dealing subsystem. System network stage. Note that the processes Head Dealer, Other-banks and Interest-rates are considered to be external to the system and are not shown.

rather than the Dealer and Clerk, and hence the datastreams pld and rdp have been changed. There were no actions in Dealer to partition out individual deposits (necessary to update Deposit state vectors) and consequently this connection is better specified in the Clerk process.

Divide-customers acts as a synchronizing process for the forecasting subsystem by controlling the activity of Forecast and Select-duration. Filter processes have been added for input validation and implementing the user dialogue, with submenus for the Clerk and Dealer processes and for the user interfaces of the Customer, Set-limits, Ratings and information subsystem functions. The system network is depicted in Figures A.20 to A.22.

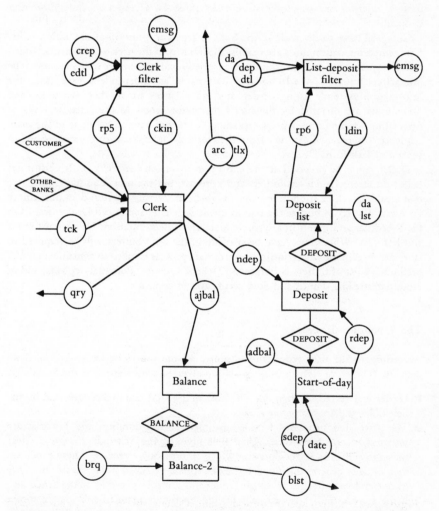

Figure A.22 System specification diagram: Deposit recording subsystem. System network stage.

A.8 IMPLEMENTATION STRATEGY (IMPLEMENTATION STAGE)

Scheduling starts by inverting input processes with respect to the Scheduler. The main input datastream is fnd, hence the first process to be inverted as a subroutine of the Scheduler is the filter process for Customer. The usual heuristic of inverting the nearest neighbours in the SSD sequence could be followed to give an inversion hierarchy of Customer calling Select-duration followed by Forecast, Divide-customers, Dealer and Clerk which finally calls Balance and Deposit. But this solution would conflict with the user requirements which explicitly state that the dealers', clerks' and the forecasting tasks must be accessed separately on user demand.

Cuts will have to be made in the SSD to separate it into implementation units. Some implementation units are readily apparent from the user requirements, which state that parts of the system should be accessed separately on the user's request. The interactive function which updates Ratings is an independent user task, the information system's listing of deposits and balances are another two units, and other units are dictated by the user's desire for access to the separate tasks of forecasting, dealing and deposit recording. Cuts will have to be made in the main sequence to separate Customer from the Dealing-forecast part of the system and to separate Clerk from Dealer.

Dealer could be inverted at the end of an inversion hierarchy with Customer, Select-duration, Divide-customers and Forecast, but the user requirements dictate that Dealer has three suboptions, one of which (answering queries) does not require Forecast. To accommodate the requirements we shall have to call Dealer first from the Scheduler and invert in a separate hierarchy Select-duration, Divide-customers and Forecast. Where cuts have been made in the SSD buffers will be required to store datastream messages until the appropriate process is called to consume them. A preliminary list of buffers is Forecasts, Tickets, Queries, Potential deposits; and of input datastreams, Funds and Sold certificates of deposit.

The Scheduler process

According to the user requirements nearly all the processes are to be called on demand. Therefore eight processes will be inverted with respect to the Scheduler:

- Dealer is inverted on its own. The buffers qry, fcst and awkcst are read by the Scheduler and passed to the Dealer.
- The associated processes Divide-customers, Select-duration and Forecast are inverted in one hierarchy. When first invoked the Scheduler passes a Clear message and calls Select-duration which in turn calls Forecast. These processes then cycle to process the input. If the first forecast is successful Forecast will write its output to a buffer ready for the Dealer and control is returned to the Scheduler. If the forecast fails Forecast calls Divide-customers again to split the customer group and then processes the 'awkward' group. Divide-customers recalls Forecast to handle the 'easy' group.

- Customer is called separately to process the receipt of money from the treasury department.
- Order-clerk calls Deposit and Balance to add new deposits.
- Start-of-day calls Deposit and Balance to remove old deposits.
- Ratings which displays and updates bank credit ratings.
- List-deposit which displays deposits.
- Balance2 which displays balances.

The system has become highly interactive, which means that it has to process all the major inputs (New money, Limits, Ratings, Strategy) and all the user requests and updates. Clearly the Scheduler is going to become too complex and unmanageable if it processes all the input, so a functional division of responsibilities will be made between the Scheduler process and filter processes:

Scheduler responsibilities

- Process user requests at the system option level.
- Call the correct process and its filter process.
- Maintain and manage buffers for all datastreams which pass data between separate programs and datastreams which enter the system across the system boundary, i.e. Ticket, Money and Queries.

Filter responsibilities

- Process user requests at the suboption menu level.
- Validate all input required for execution of the underlying process.
- Create and handle the user interface.
- Read state vectors necessary for validation.
- Call the underlying process supplying valid data and user requests.

Scheduler duties

- Input rep, date.
- Manage buffers fndgp, fcst, qry, tck.

With this functional division the Scheduler structure text becomes:

```
Scheduler Seq
    Read(rep,date)
    Scheduler Body Iter(while input)
        Main-part Sel(date rec = newday)
            Call Start-day
        Main-part Alt(rep rec)
            Good Reply Sel(rep = 1)
                Read (qry) buffer;
                Read (fcst) buffer
                Call Dealer-filter;
                Write (tck) buffer;
            Good Reply Alt(rep = 2)
                Read (tck) buffer;
```

```
              Call Clerk-filter;
              Write (qry) buffer;
          Good Reply Alt(rep = 3)
              Call Customer-filter;
              Write (fnd-gp etc.) buffer;
          Good Reply Alt(rep = 4)
              Call Select-duration;
              Write (fcst) buffer;
          Good Reply Alt(rep = 5)
              Call Limit-filter;
          Good Reply Alt(rep = 6)
              Call Rating-filter;
          Good Reply Alt(rep = 7)
              Call List-deposit-filter;
          Good Reply Alt(rep = 8)
              Call List-deposit-filter(for Balance2);
          Good Reply Alt(rep <0 or >8)
              Disp-err-msg;
          Good Reply Alt(rep = 0)
              Disp-signoff-msg;
          Good Reply End
      Scheduler Action End
    Main-part End
    Read (rep, date);
  Scheduler Body End
Scheduler End
```

The Scheduler appears to the user as the main system menu:

Deposit Dealing System

1. Dealing operations
2. Deposit recording
3. New customer deposits
4. Forecast
5. Edit customer credit limits
6. Edit bank ratings
7. Display customer deposits
8. Display customer balances

Please select option (0 to exit):

Filter duties

Filters handle all user input relevant to the chosen option and pass datastreams to the underlying process consisting of the user choices, input data supplied by the user and input datastreams. The structure text of filter processes has not been elaborated; most filters will have structures which ensure only valid option requests and data details are passed to the underlying process.

For example, the Rating filter will check a valid option has been chosen (add, delete, change or quit), accept data (Bank ID, Customer ID, Rating value), validate the IDs against state vector files and then pass the option request and the data to Rating process. The Rating filter also ensures the data is not corrupt or false, e.g. a rating of -100 will not be accepted.

Balance2 has only a small input requirement, and therefore its structure is elaborated in the information subsystem filter for Deposit-lister.

Other input is assumed not to require validation – e.g. the sold certificates of deposit (sdep) which are transferred electronically from the treasury system.

- *Dealer filter*. Inputs: drq, hd, qry, fcst (the latter two from the Scheduler buffer).
- *Customer filter*. Inputs: fnd from the input buffer.
- *Clerk filter*. Inputs: crep, edtl, tck (the latter input from the Scheduler buffer).
- *Rating filter*. Inputs: ra, rdtl.
- *List-deposit filter*. Inputs: da, depdtl.

The filters will appear to the user as submenus for dealing operations and for editing limits and ratings, e.g.:

Customer Credit Limits Submenu
1. Display limit
2. Add new credit limit
3. Change existing limit
4. Delete existing limit
Please select option (0 to exit):

Files and buffers

State vector separation creates files for the following processes:

- *Deposit*. Holds information describing individual customer deposits.
- *Balance*. Contains the balance with a particular bank for one customer. Not all the possible combinations of this state vector will be present; for instance, some customers may never have balances with bank X, while other customers may hold balances with all the banks (although this would be unusual).
- *Limit*. Expresses a customer's credit limit or complete ban with another bank. Like the Balance state vector it may be present for only a few of the total number of banks.
- *Rating*. Holds the FNBR's rating of other banks. There is an occurrence of this state vector for every other bank with which deposits are placed.

Other state vectors are external to the system: Date, Other-bank, and Interest-rates.

The following buffers are created to store data at points in the SSD where cuts have been made, breaking the connection between processes:

- *Money-group*. Stores money which has been grouped to form a potential deposit before it is passed to the Forecast process.
- *Cust-funds*. Stores the duplicate potential deposit while the Forecast process runs. This buffer is emptied irrespective of whether the forecast was successful.
- *Forecast*. Stores the forecast list until required by the Dealer.
- *Ticket*. Holds tickets created by the Dealer until the Clerk is called to process them.
- *Query-tickets*. Holds queries created by the Clerk until the Dealer is ready for them.

Two input buffers are required to store data until the Money or Start-of-day processes are called to consume the input Sold-CDs or New-funds.

The implementation design of the system is illustrated in Figures A.23 and A.24.

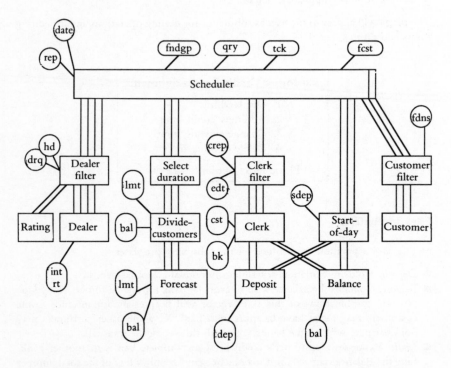

Figure A.23 System implementation diagram 1: Deposit dealing system. The true number of datastreams in the inversion hierarchy has not been shown.

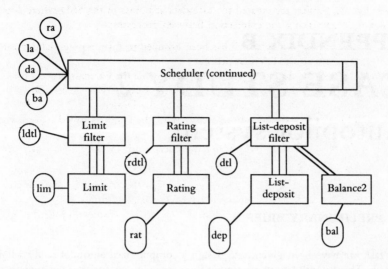

Figure A.24 System implementation diagram 2: Deposit dealing system. Scheduling the Information subsystem.

APPENDIX B

CASE STUDY 2
Autopilot system

B.1 PRELIMINARY BRIEF

The task we have been given is to design a computerized autopilot to fly a light aircraft. The autopilot is to be simple in comparison to systems on modern jet airliners; its job is just to fly a straight course as specified by the pilot. To fly an aeroplane we need to control three things.

- *The height of the aircraft*, which is altered by the elevators on the tail. If the elevators are turned upwards the plane climbs, while turning them downwards causes the plane to dive.
- *The direction*, controlled by the rudder, which can be turned either left or right to change direction.
- *The engine throttle*, which changes the speed of the aircraft. If the engine speed is allowed to fall too low the plane will stall and fall out of the sky – clearly a situation to be avoided.

We are provided with some input devices which will tell us if the plane has changed its course or height.

- A *magnetic compass* which signals changes in direction to the right or left from a preset bearing. Thus if the compass is set to 180 degrees (south), it will send a signal when the aircraft drifts either to the right or to the left.
- An *altimeter* which checks the height of the aircraft against a preset level; if the aircraft hits turbulence the change in height will be detected and signalled to the autopilot.
- An *airspeed indicator* which measures the speed of the aircraft relative to the air surrounding it.

Transponders are fitted to these devices which allow them to be calibrated to a particular reading. The transponder monitors output from the instrument and sends a signal which codes any deviation from the set reading. The code is set in a bit pattern which is adjusted according to the scale of each instrument. For the altimeter it is set at units of 10 metres and the resulting code in an eight-bit byte is:

11111111 no deviation from height
10000001 deviation of 10 metres higher than setting
10000011 deviation of 20 metres higher than setting
00000001 deviation of 10 metres below setting
etc.

The code has the first bit set as an up/down flag and the next seven bits can signal 128 variations of instrument reading. The transponder signals regularly ten times per second. Each instrument is calibrated in a similar way: the compass transponder signals deviation in half-degrees to the left or right from a compass bearing; and the airspeed indicator signals deviation in knots (the aeronautical equivalent of miles per hour) faster or slower than the set speed.

To control the aircraft there are three mechanisms operated by servo controls: the rudder which controls direction, the elevator which controls height and the engine throttle which alters the engine revs and hence the speed. Two mechanisms respond only when they are given a signal; once the signal terminates the rudder and elevators return to true. In contrast the throttle responds by changing its setting by deflecting the throttle lever and maintaining the new position as its set level. All the mechanisms are controlled by a similar code to the transponder code.

Hence as long as the code 00000001 is sent to the rudder it will be deflected one degree to the right. The rudder will remain in this position only if the signal continues; if it does so the aircraft will turn gradually to the right. As soon as the signal stops the rudder returns to true and the turning will cease.

The brief for the computer system is to develop an autopilot which flies the plane straight and true at the same speed. When the pilot engages the autopilot readings are to be taken from all the instruments and used as settings. The autopilot must then maintain those settings by monitoring instruments and taking correcting actions with the aircraft's controls. Deviations from autopilot settings may be caused when the aircraft encounters either sidewinds which blow it off course or air pockets which cause it to lose height.

These disturbances are assumed to be temporary, apart from headwinds which can cause the aircraft to slow down and tailwinds which will make it fly faster. Our user has asked us to take correcting action in this case and reset the throttle level to take account of head and tail winds.

The pilot can alter the setting of the compass while the autopilot is engaged but no other settings can be changed. When the autopilot is disengaged the link between instrument and aircraft controls must be discontinued.

B.2 INITIAL ANALYSIS: ENTITIES AND ACTIONS (MODELLING STAGE)

To start, we have to identify possible entities in this system. The objects which have to be modelled in the real world are the instruments which tell us how the aircraft is flying. These objects will send events to the system to which it will have to respond. We should also model the objects with which our system will interact. These are the control surfaces and the pilot.

The major activity in the system is controlling the aircraft, which has to be carried out by the autopilot. So Autopilot is another candidate as an entity with a life history of controlling actions. Connected to the autopilot there are sensors (Altimeter, Compass and Airspeed) and effectors (Rudder, Elevator and Throttle). Although these devices are provided for us we still need to process their signals so we can make sense of them. Provisionally we shall allocate entities to these six mechanisms. The Pilot will have some actions we need to consider although most of the activity of manual piloting need not concern us.

On a first look the main system problem appears to be about maintaining a steady state with the autopilot, so we could start by modelling the actions which are required to control the aircraft. The external events which will become action attributes are changes in height, deviation of course, changes in airspeed, engage/ disengage autopilot and change course. Preliminary analysis gives us the following list of actions.

Entity: Autopilot

Climb. Climb higher than current altitude.
Dive. Go below current altitude.
Slow. Slow airspeed by reducing throttle compared with the setting.
Speed up. Increase airspeed by increasing throttle more than the setting.
Turn left. Alter course to the left.
Turn right. Alter course to the right.

The attributes for the actions will be similar, e.g.

Climb. Direction for elevator (up), degrees deflection from true.
Turn left. Direction of turn, degrees of turn.
etc.

All these action attributes have a code for the type of change to the control surface and a code for the amount of change. The timing of change is controlled by how long the event continues, in other words how many times the message is repeated. The actions could interact; for instance, when the plane dives it will naturally speed up and the converse when it climbs. This entity will also need actions to respond to the pilot's wishes and some way of receiving signals from the instruments, e.g.:

Start-session. Autopilot engaged.
Change course. Pilot changes compass bearing.
End-session. Pilot disengages autopilot.
Monitor. Monitor incoming signals from instruments.

ACTION ATTRIBUTES

Start-session. Time, session flag.
Climb/Dive. Direction, number of degrees.
Slow/Speed-up. Change in airspeed.
Turn-left/right. Direction, number of degrees.
Change course. Direction, number of degrees.
End-session. Time, session flag.
Monitor. Signal type, quantity.

At any time these attributes must describe any change of state in the outside world.

The entity attributes will be updated by the Autopilot actions and describe its state. A provisional list is:

current course	height
airspeed setting	engaged/disengaged
session start time	session duration

Next we consider the Pilot entity. There are only three things the pilot can do which are relevant to the system.

Entity: Pilot

Engage. Switches autopilot on.
Disengage. Switches autopilot off.
Change course. Alters compass bearing.

The attributes for these actions will be a simple switch (on/off flag) for engage/ disengage and direction, and degrees change for change course. In this entity the entity and action attributes will be the same.

Entity: Altimeter (and Airspeed)

The sensors/transponders may be expected to have similar actions. First they must set the reading when they receive a signal from the autopilot; then they have to monitor deviation from the preset signal, which can be one of three conditions: no change, + change or − change depending on which way it is viewed. At the end of a session the transponder has to disengage itself.

The Altimeter actions are:

Set height. The altimeter reading when the engage signal is received is taken as the reference level.
Compare. Compare the altimeter reading with the preset level.
Lower. Signal height has been lost.
Higher. Signal height has been gained.
Same. No change from preset value.

ACTION ATTRIBUTES

Set height. Metres altitude.
Compare. Set height, new height.
Lower. Metres lower, difference flag.
Higher. Metres higher, difference flag.
Same. Metres altitude.

Similar action definitions can be prepared for Airspeed (with the difference of slower/faster).

For Compass a further action is required to detect a possible change of direction by the pilot. Otherwise the action list is similar to the Altimeter's.

Entity: Compass

Set course. The compass bearing when the engage signal is received is taken as the course.

Measure. Compare the compass bearing with the course.

Move-left. Deviation to the left.

Move-right. Deviation to the right.

On-course. No change.

Poss-change. Look for course change signalled by the pilot.

ACTION ATTRIBUTES

Set course. Compass bearing in degrees.

Move. Difference flag, degrees difference.

etc.

So far we have no actions to compensate for changes in airspeed which result from headwinds and tailwinds. We could add some actions in Autopilot to react to changes and calculate the required compensation in speed, but as yet we have no details of how and when to compensate for headwinds. Some algorithm will have to be designed; this is probably going to be an interactive function so we should not include further details at this stage.

B.3 ENTITY STRUCTURES AND THE INITIAL MODEL (NETWORK STAGE)

The initial model will have five processes: Pilot, Autopilot, Compass, Altimeter and Airspeed. The mechanisms which control the rudder, elevator and engine speed are not within the system as these are standard parts of the aircraft's design. For the time being we have chosen not to include the process to correct for changes in headwinds – we will add this process later in the network stage when the system specification diagram is elaborated.

Model process: Pilot

This process models the events created by the real world pilot. It has a life history which starts with the real world pilot engaging the Autopilot and ends when the pilot disengages it (Figure B.1). In between there may be zero or more course changes but no other actions by the pilot are specified so we only need to model a 'Poss course change action'.

Model process: Autopilot

Autopilot (see Figure B.2) models the responses to external events by the autopilot mechanism. As soon as it is engaged the autopilot must start by receiving the settings of course, height and speed which it should maintain. During autopilot control there is considerable uncertainty. At any time one of the three variables (height, course, speed) may change, necessitating corrective action. Also the autopilot must be receptive to the pilot's commands, which may be to change course or disengage the autopilot.

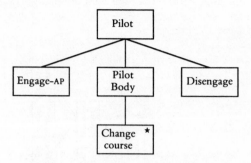

Figure B.1 Model process: Pilot.

The process could be modelled as a series of autopilot sessions which contain a series of possible manoeuvres terminating in a disengage, but the problem is that we don't know when the pilot will disengage the autopilot. In view of this run time uncertainty backtracking seems appropriate. This theme could be taken further using a nested backtracking structure, the first level being the Autopilot session. Positing it continues until the pilot Quits from the session, and the second level Positing no deviation from course Admitting there was and taking corrective action. However, this analysis would leave the second Posit branch with very little to do except monitor the instruments. We choose to model the process with only one layer of backtracking to reduce the complexity of the process specification.

Autopilot starts by getting the course settings from the instruments after receiving an engage signal from the pilot. The main process body is a Posit while the session continues and an Admit that it has finished (due to the disengage signal) and resetting the instruments as a result. Within the Posit branch the autopilot has to monitor the instrument readings continually and take corrective action if necessary. This is modelled by a series of Poss-change actions. The order of monitoring does not appear to be important since all the instruments will be scanned several times a second.

Within the Posit branch we need to monitor for the disengage signal. This may happen at any time, so Quit probes are inserted after each monitoring action. Unfortunately two different signals may be received from the pilot, either a disengage or a course change. For the first we need to reset the instruments and finish the session, but for the second we need to reset the compass and continue the session. The choice is to model this clash with nested backtracking or to use a possible action for the course change. Having decided to keep to one layer of backtracking, we shall keep the course reset actions within the Posit branch.

Model process: Altimeter

Altimeter (Figure B.3) starts by setting the height to be maintained when it receives a start signal from the autopilot. It then has to monitor any change of height compared with the reference height until a stop signal is received. The main body is

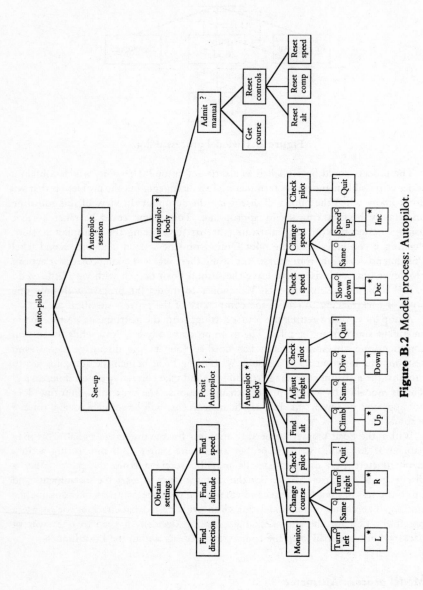

Figure B.2 Model process: Autopilot.

an iteration of measurements which may be either higher, lower or signal no change.

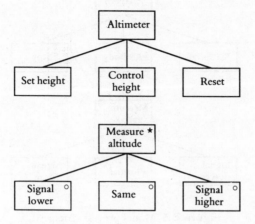

Figure B.3 Model process: Altimeter.

Model processes: Compass, Airspeed

These are similar in structure to Altimeter and consist of an iteration of readings against a set level followed by a stop event (see Figures B.4 and B.5).

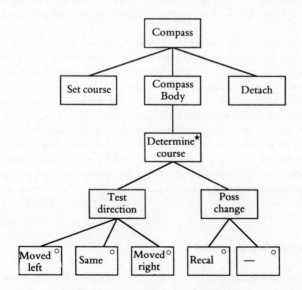

Figure B.4 Model process: Compass.

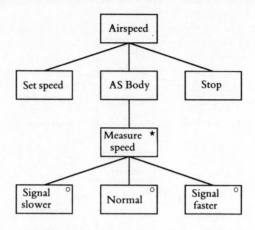

Figure B.5 Model process: Airspeed.

Further modelling problems

So far we have modelled the part of the system which creates event input if the aircraft departs from its set course. Another model has been created of the possible combination of events which may happen to the aircraft during flight, which is described in the autopilot. The control surfaces of the aircraft have not been modelled, however.

Two of these surfaces, the rudder and elevator, can be considered to be outside the system boundary because their properties and the mechanisms which control them are well known. The engine throttle requires more attention. It has a series of settings at which it can be fixed when the autopilot is engaged, whereas by contrast the other control surfaces will only depart from the resting position when signalled to do so and automatically return to a resting position. The throttle needs to set the engine revolutions to a level which matches the speed when the autopilot was engaged and then has to manage the engine revs to maintain that speed. In view of this a throttle control process is required. This will be an interactive function which will create input for other model processes. The interactive functions are omitted from the initial model but will be added to the first draft of the system specification later in the network stage.

System network: initial model

The model processes which model instruments (Compass, Altimeter and Airspeed) receive input datastreams from the Autopilot to give them their initial settings. These processes monitor the corresponding real world instruments with state vector connections because they require quick snapshot readings. The Pilot model process will receive datastream input as the pilot's commands.

The Autopilot will inspect the state vectors of Compass, Altimeter and Airspeed to find changes in the instrument readings and receives datastream input from the Pilot process communicating engage and course change signals. State vector connections are used to inspect the instrument processes because regular snapshot views of instrument readings are required. Autopilot outputs datastream messages to the flight control processes (trn, vrt) and to the engine (rev).

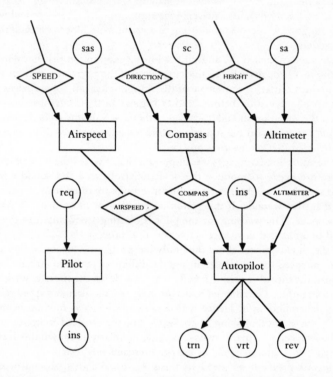

Figure B.6 Initial model: Autopilot system. Note that the instrument model processes monitor events in the transponders which are outside the system by state vector inspection.

B.4 NOTES ON FURTHER ANALYSIS

The results of our analysis so far are shown to the chief engineer who points out a few changes that will have to be made. The first problem is that some of the flight variables interact. When the aircraft dives it increases airspeed and conversely when it climbs it will decrease airspeed. But how do we know whether we have lost airspeed because of a climb/dive or because of another factor such as an increase in headwind? Maybe we could accommodate this by putting in an adjustment for the throttle whenever we climb or dive the aircraft. When we ask how much the airspeed would decrease in a climb of a set angle, the chief engineer informs us that the decrease in airspeed when climbing depends on how steep the climb is, the airspeed and how heavily loaded the aircraft is. We cannot measure all these factors, so instead we settle for monitoring the airspeed for any change when climbing or diving. The aircraft designer informs us that we need take no action unless the aircraft is in danger of stalling (the speed decreases to a level at which it can no longer fly), in which case the autopilot must disengage and warn the pilot. This suggests we need a monitoring process in Autopilot to detect if there is a danger of stalling.

So far we have modelled the autopilot with the assumption that nothing can go wrong. But what happens if the engine cuts out or if the aircraft encounters extreme turbulence which buffets it so much that the autopilot has not got sufficient time to recover from one deviation before another begins? In these circumstances we are advised that the autopilot should disengage, otherwise valuable seconds may be lost while the pilot switches off the autopilot before regaining control. This requirement could be included in a monitoring process.

Before the autopilot disengages it is important that it warns the pilot of its action, and it therefore needs to monitor the deviation from true and sound a warning buzzer and switch on a light if the instrument settings are more than 30 per cent out for more than five seconds. If this condition persists for more than ten seconds it should disengage. The warning for the pilot is creating a new input to the system and should be modelled as an interactive function process.

We ask the chief engineer about the headwind problem. Our initial solution was to check the airspeed for any deviation and if a difference persisted for a set length of time to adjust the throttle to either speed up or slow down the aircraft. We learn that life is not that simple. The airspeed indicator measures the aircraft's speed relative to the air mass surrounding it. Hence with a tailwind the air is moving in the same direction as the aircraft, making it go faster, but the airspeed indicator will still measure the airspeed as if the air mass was static. Similarly a headwind will slow the groundspeed of the aircraft while its airspeed remains unaffected.

To solve our problem we need to measure the aircraft's groundspeed. This is done, we are told, by a radio altimeter which measures the time difference between transmitting and receiving a radio wave bounced off the ground below. The time difference is a function of the aircraft's height and the speed at which it is travelling. We will need to access the radio altimeter at set intervals to check the groundspeed against the airspeed. If we find a difference then the aircraft has encountered either a headwind or a tailwind and compensatory action will have to be taken. To deal with this problem another interacting function is needed to monitor changes in airspeed and groundspeed. In this case it needs to monitor not the Autopilot but another model process which so far has been omitted from the specification. This model process will describe the event history of another instrument, the radio altimeter.

B.5 SYSTEM SPECIFICATION DIAGRAM (NETWORK STAGE)

We are now in a position to add interactive functions to our model processes and complete the first draft of the system network. There are three interactive functions to add: one for the throttle, another to monitor for violent changes in height and direction which may trigger a premature autopilot disengage, and the other to correct any changes in airspeed due to headwinds etc. We shall also elaborate the Autopilot model process to add new details about its life history. These describe further events which may lead to premature termination of its life history, such as mechanical failure or violent turbulence. Another addition is a model process to describe another instrument, the radio altimeter. This should have been included in the initial model but owing to an analysis error we were unaware of its existence.

At this stage the information function to warn the pilot of an impending disengage will not be added. The complete user requirements have to be analysed before we proceed to specify the output subsystem.

Interactive function: Throttle

Throttle (see Figure B.7) starts by receiving the setting signal from Autopilot (Acquire speed) and then sets the engine revs to an appropriate value using a simple algorithm to relate engine revs to the expected airspeed (revs × load factor = exp. airspeed). A calibration is necessary so that Throttle can determine how much to change the engine revs according to differences in airspeed. The rev counter sends signals to Throttle in a similar manner to the other instruments' transponders. Throttle then has an iteration of possible changes in engine revs which depend on differences in airspeed. Airspeed is monitored by Autopilot inspecting the state vector and if a change is detected Autopilot writes a datastream (sp) to Throttle

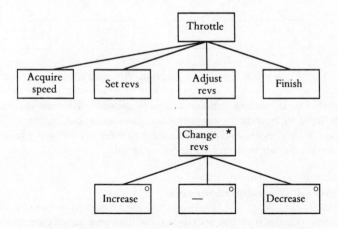

Figure B.7 Interactive function: Throttle.

which causes it to increase or decrease the engine rpm by an output datastream (rev). This continues until Throttle receives the disengage signal from Autopilot.

Interactive function: Airspeed corrector

This function (see Figure B.8) will have to sample the state vector of the groundspeed model processes. An initial reading of groundspeed will be taken when the autopilot is engaged. Readings are then taken at regular intervals to determine if there is any difference between the initial and new groundspeed readings. The chief engineer informs us that reading need only be taken at one-minute intervals and that the system should react only if there is a deviation from the initial setting of 5 per cent or more.

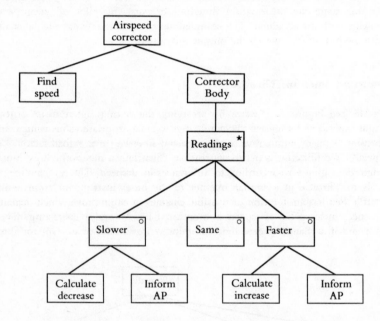

Figure B.8 Interactive function: Airspeed corrector.

If the change in readings shows a slower groundspeed than expected then messages have to be sent to the autopilot to increase the engine revolutions. Conversely if a higher groundspeed is found then there is a following tailwind and the engine throttle can be decreased to reduce speed.

Model process: Groundspeed monitor

This process (Figure B.9) has two inputs. A series of radio pulses is directed vertically downwards and bounced back off the ground below. The time delay between the transmission of one pulse and the arrival of its echo is a function of the transmission velocity of radio waves, the height of the aircraft and the distance it has travelled between transmitting and receiving. The speed of the aircraft can be calculated from the time interval between the radio beam being transmitted and received back again. The model process uses the height setting from the altimeter state vector to calibrate itself and calculates the groundspeed from the radio pulse time interval.

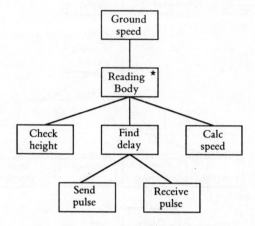

Figure B.9 Model process: Groundspeed monitor.

Interactive function: Turbulence monitor

This function (shown in Figure B.10) has to inspect the Autopilot model process's state vector. If changes in the instrument readings, as analysed by the Autopilot, vary by more than 30 per cent from the setting for more than ten seconds then the Turbulence monitor writes a disengage message to the Autopilot. The scan frequency of the Autopilot state vector is set at once per second. The process maintains a stack of ten readings in a FIFO buffer; the manage stack action adjusts the queue for the next reading.

Modifications to the Autopilot model process

New actions are added as Quits to test for the additional causes of a premature termination of the Autopilot's life. These are disengage messages from the Turbulence monitor, Engine failure messages from outside the system boundary, and airspeed which may indicate stalling. The adjust speed action now uses groundspeed as its measure.

Another datastream Read will have to be added for messages from the airspeed corrector. The admit branch is elaborated to deal with emergency terminations (warn disengage) and more normal pilot-initiated disengages (see Figure B.11).

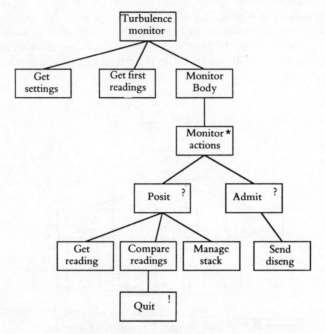

Figure B.10 Turbulence monitor.

System network

The choice of connections in the system depends on how long-term a view is required. Autopilot output and input from interactive functions will be event streams which persist for a potentially long time, and consequently datastreams are used.

Messages have to be passed between the Autopilot and Pilot processes and the Autopilot has to scan the instruments. For most of the input connections from the instrument model processes to the Autopilot a snapshot is required of their readings, and hence state vector connections are used. When scanning instruments the Autopilot only needs to know when a change has occurred, which suggests a state vector connection rather than a datastream. Use of a datastream could cause blocking of the Autopilot process if a message was absent, a consequence that is undesirable even for a short time.

The instruments – Compass, Altimeter and Airspeed – are connected by state vectors to Autopilot which monitors them for any changes. In addition they have datastream connections (sc, sa, sas) with Autopilot for the 'start and finish' signals.

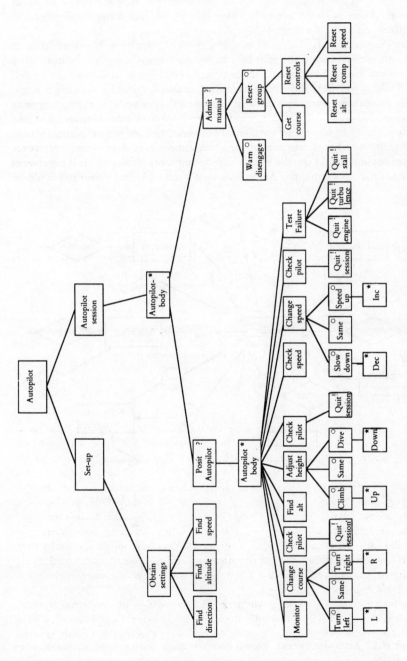

Figure B.11 Model process: Autopilot (2nd version after addition of extra quits).

These event signals must be read for the instruments to start and stop a session of readings. There is no danger of blocking the system with a start-up message, so datastream connects are used.

The Pilot process is connected to the real world pilot by a datastream connect (req). All messages from the pilot have to be obeyed, and therefore the mandatory Write–Read connection of a datastream is used. The Pilot process transmits the real world pilot's actions to the Autopilot using a datastream (ins) for the same reasons.

The datastreams trn and vrt are used to connect the Autopilot to the flight control processes, Elevators and Rudder, which are outside the system. Messages from the Autopilot must be obeyed by the control processes, and hence datastreams are used because the message stream persists for a long time period. Use of state vectors for this connection would run the risk of introducing more timing errors if the control processes had to sample the Autopilot's state vector. A vital command could be

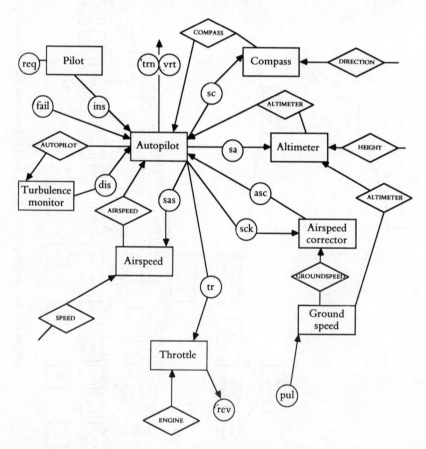

Figure B.12 Autopilot system: system network stage. System specification diagram (first draft).

missed because the sampling speed was too slow or too fast, and in addition the Autopilot wants its instructions carried out in strict sequence. The FIFO nature of a datastream ensures that the sequence of instructions is preserved. The Throttle process is connected to the Autopilot process using the datastream tr for similar reasons. It then communicates to the extra-system process Engine with a datastream.

The first draft of the system network showing model processes and interactive functions is shown in Figure B.12.

B.6 USER REQUIREMENTS ANALYSIS

So far our autopilot has only had to communicate with the aircraft controls, but flight safety regulations dictate that autopilot instructions have to be recorded on the aircraft's black box flight recorder. The black box is a simple device which is essentially a multiple-track tape recorder. We have to program the message stream for the tape which will record what happened to the controls during a flight. One track will be allocated to the rudder, one to the elevators and so on. The regulations require a coded representation of the autopilot instructions to the control surfaces with a time stamp accurate to 0.1 seconds. In the interests of economy we need store information only when something happens, i.e. when correcting instructions are issued by the autopilot. There are separate recordings of the cockpit instrument readings which are not part of our system.

A second user requirement is given by the chief engineer, who has decided that the original specification did not go far enough. One of the problems with autopilots is a slow drift off course. A persistent sidewind could cause the aircraft to deviate slightly off course even though the autopilot is taking corrective action. This is because of the time delay between the instrument detecting the deviation from true and the flight controls correcting the aircraft's direction. During that time the aircraft may have been blown sideways, but when the autopilot has corrected the heading the sideways drift remains undetected as the aircraft is now on the same course although in reality on a parallel track displaced laterally.

To counter this problem the system should warn the pilot when drift may be occurring so he can check his navigation and if necessary reset course. The critical factor to monitor is the relative number of rudder corrections in one direction. If there are several corrections in the same direction then there is a significant chance of drift. The system should monitor corrections in both directions and if ten more corrections are made in one direction than in the other then a warning message is to be displayed for the pilot.

A final requirement is for the autopilot to give a display of the current course, height and airspeed settings so the pilot can check these values. The figures will be displayed on a small LED device on top of the autopilot. Our system has to feed it with information in ASCII code and update the information when it changes. This will also apply to possible course corrections made by the pilot.

Our brief is to change the system to incorporate the drift monitoring, the flight recorder and the control settings display.

B.7 SYSTEM NETWORK STAGE: OUTPUT FUNCTIONS

The user requirements necessitate addition of two new processes to the system, because actions have to be added to fulfil the specification and JSD prohibits adding new actions inside existing model processes. Imposed functions will be added for the black box recording and for the drift control. The third requirement for feeding the LED displays is simpler and may be satisfied by simple embedded functions. In this system there is little need for an input subsystem, as we are assured that messages from the instruments are reliable and that the pilot cannot send an incorrect message as the only switches on the autopilot box are an engage/disengage toggle and the course change dial.

Imposed function: black box

This process (Figure B.13) receives a stream of messages from the Autopilot which describes every manoeuvre. Messages are written by the Change-course, Adjust-height, Change-speed and Check-pilot actions to record all the changes brought about by the autopilot. Datastream connections are used because recording the events in sequence is important. The datastream rc will contain a heterogeneous collection of message types, but that is of little importance to the black box which has the simple job of time-stamping the message and then recording it.

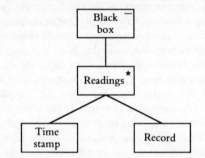

Figure B.13 Function process: Black-box.

Black-box is an iteration of time stamps and records. This sequence is perpetually active as the box is never switched off; when the autopilot is disengaged it continues to monitor the pilot's actions using another part of the system which is outside the terms of reference of this system.

Imposed function: drift monitor

Drift monitor (see Figure B.14) has to record the number of turns being made. It is connected by a datastream crs written by the Change-course action in Autopilot. The datastream messages may come in groups of Turn lefts or Turn rights. Drift monitor is only interested in a single turn, not in how long the rudder was maintained in that position (remember this was effected by continuous messages) so Detect-course has to read the messages until either there is a time gap with no crs messages (one second) or the type of message changes (i.e. from Turn left to Turn right). This is achieved by a rough merge of the crs datastream and a TGM.

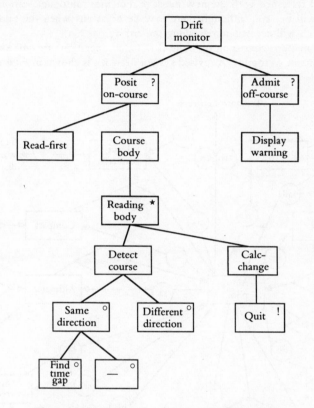

Figure B.14 Interactive function: Drift monitor.

Drift monitor has to respond to an uncertain world, i.e. whether the aircraft will or will not drift off course. Bearing uncertainty in mind backtracking is used, with a Posit that no drift occurs but Quitting if excessive drift happens. The Admit branch displays the warning to the pilot to take manual corrective action. Detect-course increments counters for turns to the left and right, Calculate-changes adds up the relative number of changes which have been made then decides if drift has occurred (if Left − Right = 10 or Right − Left = 10 then Quit to Off-course).

Embedded functions

The requirement to show the course, height and airspeed settings on LED displays can be satisfied by simple embedded functions within Autopilot. No processing of data is required, so transmission of simple messages in ASCII characters to the LED devices will suffice. The device internals need not concern our system; each LED display is designed so that it interprets each ASCII character and shows the appropriate pattern on the display. When a new message arrives the display is cleared and refreshed with the new message. For most autopilot sessions no new messages will be sent, although there may be occasions when the pilot changes course which will necessitate refreshing the display.

Simple functions consisting of Write-datastream operations are embedded in the Reset actions in Autopilot. A revised system network is shown in Figure B.15.

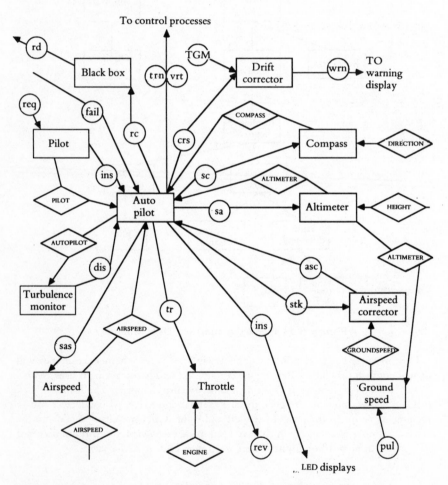

Figure B.15 Autopilot system: system network – function step. System specification diagram.

B.8 IMPLEMENTATION

A simple strategy would be to implement each process on a separate microprocessor. This does not seem unreasonable as most of the processes have to be running concurrently. The Autopilot will have to be active for long periods of time and each process that monitors an instrument will also have to be active to feed the Autopilot with information. Unfortunately we are limited by the hardware available which has been dictated by the cost accountants. They have told us that only three standard microprocessors can be used.

Implementation strategy

In view of the processor constraint time-slicing will have to be employed to give each process a share of processor time sufficient for it to complete its tasks. One processor will have to be allocated to the Autopilot which it is essential to keep running as continuously as possible. Another processor should be able to manage the four instrument-monitoring processes Compass, Airspeed, Groundspeed and Altimeter, which are all fairly small programs, and the third will have to be shared between the imposed functions Black-box and Drift corrector and the interactive functions Throttle, Turbulence monitor and Airspeed checker. The interactive functions will require a priority of this processor's resources to ensure that they have enough processor time to respond to all the input events and finish their life cycle. The imposed functions are not so time-critical, so some unequal resource-sharing may be tolerated. This still leaves the Pilot process unaccounted for.

Pilot and Autopilot do not need to be active at the same time, apart from the need for Pilot to determine whether either a course change or a disengage autopilot event has been signalled by the human pilot. We could get Autopilot to fulfil this task by directly inspecting the state vectors of switches for changing course and engaging the autopilot.

Implementation 1: Autopilot and Pilot processes

The Pilot process is dismembered, transferring the Course-change action to the Autopilot. The Pilot process becomes a simple scheduler. Autopilot is inverted with respect to Pilot which calls it when an Engage message is received. Autopilot will then start a session of controlling the aircraft and monitoring the status of the course selector and disengage switches using state vector connections. Course changes do not cause the Autopilot to cease operation; another Posit sequence of control corrections is started with the new course.

If Autopilot encounters a Disengage on the state vector then it ceases operation and returns control to Pilot, which waits until the next Engage event. The Pilot process is the scheduler on this processor (see Figure B.16).

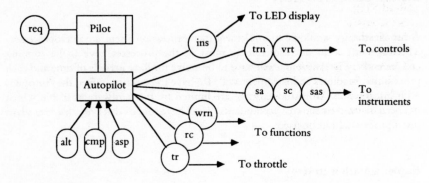

Figure B.16 Autopilot system implementation, system implementation diagram: Processor one.

Implementation 2: Compass, Altimeter, Groundspeed and Airspeed

These processes are equally important and need to be allocated one-third of the processor's resource on a time-sliced basis. Tests show that the processes, which have a similar structure, execute in a time range of 70–100 milliseconds. A clock could be used to supply time markers allocating each process the upper range value of 100 milliseconds to complete its cycle, but this may be complicated if a process runs more quickly and completes its cycle in 80 milliseconds and then starts over again. In this case it may be swapped out by the clock in the middle of a cycle.

We could control the processes with state vectors to record which part of its life history a process had reached on termination. The process could be made re-entrant, i.e. operation of the process code re-enters at the point at which it left off. The entry–exit point is coded in the state vector. This solution is unsatisfactory, however, because it may lead to an unfair distribution of processing cycles between the instruments if one runs faster than the others. The Autopilot could get more readings but these would be of little use unless a process had completed its cycle and produced a complete update of its state vector.

To control time allocation the scheduler process will have to monitor process execution time, and if a process completes more quickly than normal then the extra time must be taken up with a waiting period. This could be controlled by running the Compass, Altimeter, Groundspeed and Airspeed processes so that they execute just one cycle per call, becoming effectively a very fast batch process. Some dismembering will be necessary to relocate the Set actions of the three instrument processes within the Instrument-scheduler. The Instrument-scheduler will have to assume responsibility for checking the Start-datastream message from Autopilot and monitoring the state of the Autopilot state vector in case there is a course change. The three datastreams which switched on the instrument processes (sa, sc and sas) can be merged into one single datastream (sti) which starts the Instrument-scheduler. A possible structure text for the Instrument-scheduler is:

```
Instrument-scheduler Seq
   Read(STI);
   Set Course;
   Set Height;
   Set Speed;
   Clock-Body Iter(while APsv = On)
      Posit Reading-Cycle
         Read TGM(System clock);
         Altimeter-Seq
            Call Altimeter;
            Measure-Time Sel(100 ms not finished)
               Wait;
            Measure-Time Alt(100 ms up)
               Continue;
            Measure-Time End
         Altimeter End
         Read TGM(System clock);
         Compass-Seq
            Call Compass;
            Measure-Time Sel

               common subtree (wait)
         Compass-End
         Read TGM(System clock);
         Airspeed-Seq
            Call Airspeed;
            Measure-Time Sel

               common subtree (wait)
         Airspeed-End
         Read TGM(System clock);
         Groundspeed Seq
            Call groundspeed;
            Measure-Time Sel

               common subtree (wait)
         Groundspeed End
         Check AP
            Get SV(autopilot);
            Quit Reading-Cycle if SV = finish;
         Check AP End
      Posit Reading-Cycle End
      Admit Finish Cycle
         -
      Admit Finish Cycle End
   Clock-Body End
Instrument-Scheduler End
```

The Instrument-scheduler is a backtracking structure which is triggered by a Start instruction from Autopilot on the sti datastream. The scheduler then cycles, calling the instrument processes in turn and inspecting the Autopilot's state vector to make sure it hasn't changed to a stop state (see Figure B.17). If it has, the scheduler Quits from the cycle and waits for the next sti message. The slight inefficiency of time wasting if processes execute in under 100 milliseconds is tolerable.

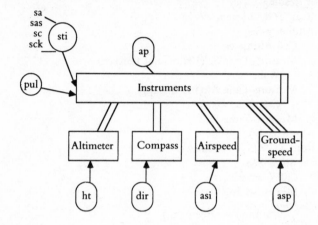

Figure B.17 Autopilot system implementation, system implementation diagram: Processor two.

Implementation 3: interactive and imposed functions

These processes are triggered by datastream messages (rc, crs, tr and sck) from the Autopilot. As with the Instrument-scheduler the datastreams can be merged. One stream will have messages for Black-box and the other stream will have two message types, one for setting the throttle and one starting the drift corrector. Black-box will be the most frequent recipient of messages, as it records all control movements by the Autopilot, and hence it requires a message stream of its own. This process, fortunately, is very simple and executes in a short time period which means it is unlikely to block operation of other processes.

The scheduler for this implementation subsystem will have to inspect the input datastreams and call Black-box automatically for all cycles. If there is a course correction, Drift corrector will have to be called, and for a speed change Throttle is called. Turbulence monitor has to execute frequently, and therefore it is given a time slice of its own; however, the Airspeed corrector is required to cycle infrequently (once a minute) and therefore it is assigned the lowest priority. The system implementation diagram is shown in Figure B.18.

Estimates of process execution time indicate that there is insufficient processor time for all five processes to execute on an even sharing of time-slicing. In view of this the processor time is divided into three parts. Black-box is allocated one time slice, Drift corrector and Throttle share another, depending on which input

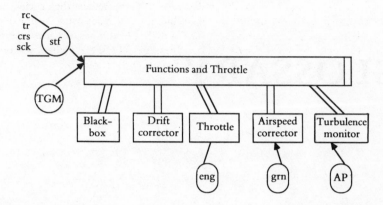

Figure B.18 Autopilot system implementation, system implementation diagram: Processor three.

messages are present, and Airspeed corrector and Turbulence monitor share the third slice with the Airspeed corrector having low priority. The Scheduler structure text is as follows:

```
Function Scheduler-3 Seq
      Scheduler Body Iter
            Read(RC) and (STF);
            Call Black-box;
            Other Action Sel(Course-correction)
                  Call Drift-corrector;
            Other-Action Alt(Speed-change)
                  Call Throttle;
            Other-Action End
            Read TGM(clock);
            Third-Slice Sel(TGM not = 60s)
                  Call Turbulence Monitor;
            Third-Slice Alt(TGM = 60s)
                  Call Airspeed corrector;
            Third-Slice End
      Scheduler-Body End
Function Scheduler-3 End
```

The datastreams crs and sp which were used in the SSD to connect Throttle and Drift corrector to Autopilot duplicate information present in the stf datastream, and are therefore eliminated.

The SIDs for the system are illustrated in Figures B.16 to B.18.

GLOSSARY

Access path A process, usually a function process, may need to inspect several other processes' state vectors. The access path defines which processes must be communicated with and in what order. It forms an input to database design.

Action/event Something that happens which the system has to respond to. Action/events are atomic and cause a response in the system and an update to the information stored within the system.

Action attribute A message conveying details about an event into the system or an event message being passed out of the system.

Backtracking A process design technique for handling run time decision-making, i.e. where the quality of information necessary to evaluate a condition is unpredictable. The technique assumes a certain choice is correct (Posit), tests for the choice being incorrect (Quit) and if it is takes remedial action (Admit).

Batch processing A method of scheduling processes so that they do not run when their inputs are available, but instead their input is accumulated and processed together at a predetermined time in a batch file.

Buffering Datastreams are considered to have conceptually an unlimited buffer. Messages on a datastream channel could therefore accumulate for infinity. During implementation finite-sized buffers may be provided for datastreams, particularly for fixed merges. Buffers may be implemented as temporary files or in working storage.

Cardinality The number of instances of a process type, shown with bars on SSD connections (see also **Instance**).

Channel scheduling A method of implementation in which system processes are directly inverted to the scheduler and not in a deep hierarchy (see also **Scheduler**).

Common action An action in which two or more entities take part which generally has the same name. For instance, Book and Member both take part in a Lend action. Common actions indicate entities' relationships.

Conditions Conditions are specified for selections and iterations in process structure.

The conditional statement is added as an elementary condition detail to PSD diagrams. For each iteration the conditional logic should specify whether there is to be another cycle or not, i.e. the termination conditions. In selections the logic must specify the conditions for each component.

Context error An error when an input message is not in the appropriate part of the life cycle sequence of a system process. The message is therefore correct but arrives at the wrong time. This can only be detected by inspecting the system process's state vector.

Datastream A communication channel which passes event messages. It is conceptually an unbounded FIFO queue.

Elementary operation An executable operation which is added as design detail to process structures. Operations are numbered and assigned to process structure components. The operation detail is expressed in the target language or in a pseudocode.

Entity An object which is involved in a time-ordered set of actions. The time ordering of entities is modelled with entity structure diagrams which express the time-ordering constraints. Entities become model processes within the system.

Entity attribute Data items which are owned by and updated by an entity's actions. Entity attributes record the history of an entity and become its state vector.

Entity role When an object has two or more sets of time-ordered actions which are independent of each other, they are modelled with separate process structure diagrams and known as entity roles; e.g. Employee might have a contract assignment role and promotion-personnel role.

Filter process A process in the input subsystem which is responsible for validation and filtering incorrect input data. There are two types of filter: a context filter to detect event sequence errors, and simple/input filter processes which implement the user interface and detect ordinary message errors.

Fixed merge Two or more datastreams are read in a predetermined order. The absence of a message on one stream will cause the reading process to be blocked. The alternative construct is a **Rough merge** (q.v.).

Implementation The last stage of JSD development in which a concurrent system network is generally transformed into a sequentially executable system by the techniques of scheduling and process inversion.

Information function A process which produces information from the model which is output as reports and displays.

Instance A specific incarnation of a type. In JSD entities are modelled as types in the process structure diagram. However, entity instances are also modelled by the state vector which records the history of each entity instance; e.g. Account is a process type, John's account is an instance.

Interactive functions These create event input within the system, i.e. they interact with the system and form a feedback loop with system processes. These functions are often regarded as automating part of the real world so events happen inside the system rather than outside it.

Inversion Program inversion is a transformation carried out on process texts converting two concurrent, communicating processes into a main program subroutine hierarchy with a suspend and resume cycle of execution. The part of the process text (or life history) which is executed in each run is controlled by a text pointer.

Iteration A process component which has only a single son. The son may be repeated zero or more times.

Lock The operation of controlling reading in a rough merge by forcing the downstream process to read messages from the locked datastream and no other until it is unlocked.

Long-running process A process which takes a significantly long time to complete its life history. Most model processes are long-running, taking months or years to complete one life cycle. Function processes in contrast are generally shortrunning although historical reports may be long-running.

Marsupial A group of actions within an entity which have a time ordering independent of the main entity. The group of actions are linked to another (undiscovered) object which represents a repeating group within the entity. Discovery of the marsupial entity causes a new entity to be modelled for the marsupial group of actions, e.g. in Customer-order, if there are many orders, then there is an order marsupial.

Model process A process which models the life history of an entity. It is described in a process structure diagram.

Modelling stage The first stage of JSD in which the real world is described in terms of event/actions and the objects to which time-ordered sets of actions belong.

Network diagram A diagram showing the connections between processes in a system. A synonym for a system specification diagram.

Network stage The second stage in JSD in which the system specification is built up by adding interactive functions and then the input and output subsystems.

Normalization The technique of designing logically independent data structures first described by E.F. Codd. First normal form analysis is achieved by JSD marsupial process analysis; this removes repeating groups from data structures. Second and third normal form analysis are steps which may be added to JSD.

Null component Part of a selection which models the possibility that an event does not occur.

Premature termination An action in a life history which brings the life to an early end before its full course has been run. If such actions occur in many parts of a life history backtracking is generally used to clarify the process structure.

Process, Program An executable set of instructions which is specified as a structure in process structure diagrams to which design detail is added in elementary operations.

Query A state vector inspection executed by the writing (upstream) process in a controlled datastream. The state vector of the reading process is guaranteed to be up to date.

Rough merge Two or more datastreams are rough-merged if they are read by the same action in a process. The reading process must read the first message which arrives on any of the channels and has no control over the reading order.

Scheduler An additional process introduced at the implementation stage to control process execution. One or more schedulers control the system processes which are inverted to become subroutines.

Scheduling group At the beginning of the implementation stage the system network is divided up into scheduling groups according to the needs of on-line *versus* batch processing, distributed processing, etc. Each scheduling group will have to be represented in a system implementation diagram.

Selection A process component which has two or more sons, only one of which may execute each execution cycle. Which son component executes is determined by the conditional logic.

Sequence A process component which has one or more sons all of which execute in sequence, reading left to right on a process structure diagram.

System implementation diagram (SID) Diagram showing the inversion hierarchy of the scheduled system with files and buffers.

System specification diagram (SSD) Diagram depicting the system as a network of concurrent communicating processes with datastream and state vector connections.

State vector A collection of the entity attributes or data items owned by a process which record its life history.

State vector inspection A read only connection between two processes which specifies that the inspecting process may access the inspected process's internal data on demand.

Structure text An alternative form of representation for process and system specification which uses indentation to show structure, and has reserved words for process component types (Seq, Sel, Iter).

System timing This is specified in narrative form as timing of response requirements, process execution time, etc. During specification timing or process activation may be controlled by time grain markers.

Time grain markers Datastreams which contains time messages. Time grain markers (TGMs) can be used to trigger and synchronize processes.

Type A model which describes the structure and time ordering of many individuals. Many process instances are modelled as a process type (see also **Instance**).

INDEX